The Art and Science of the

CRYSTAL PALACE DINOSAURS

The Art and Science of the
CRYSTAL PALACE DINOSAURS

Mark P Witton *and* Ellinor Michel

THE CROWOOD PRESS

First published in 2022 by
The Crowood Press Ltd
Ramsbury, Marlborough
Wiltshire SN8 2HR

enquiries@crowood.com
www.crowood.com

© Mark P. Witton and Ellinor Michel 2022

All rights reserved. No part of this publication may be reproduced or transmitted in any form or by any means, electronic or mechanical, including photocopy, recording, or any information storage and retrieval system, without permission in writing from the publishers.

British Library Cataloguing-in-Publication Data
A catalogue record for this book is available from the British Library.

ISBN 978 0 7198 4049 4

Cover design by Sergey Tsvetkov
Cover photographs by Kevin Ireland; Crystal Palace Park, print by Baxter, courtesy the Wellcome Collection, CC-BY 4.0

The Friends of the Crystal Palace Dinosaurs is a charity dedicated to the conservation and interpretation of the animal sculptures, geological illustrations and landscapes of the Grade I listed historic site. All proceeds from this book go to supporting their aims. (Registered with the Charity Commission for England and Wales no.1165231)

Graphic design and layout by Peggy & Co. Design
Printed and bound in India by Parksons Graphics

Contents

PART 1	ISLANDS COVERED BY STRANGE FIGURES	7
1	Historic prehistory in South London	8
2	Ancient worlds through a Victorian lens: planning the Geological Court	21
3	Bricks, iron and tiles: rebuilding the past	39

PART 2	ANIMALS LONG SINCE EXTINCT	67
4	The sculptures: mammals	68
5	The sculptures: *Mosasaurus hoffmanni*	91
6	The sculptures: flying reptiles	96
7	The sculptures: dinosaurs	104
8	The sculptures: *'Teleosaurus chapmani'*	121
9	The sculptures: enaliosaurs	126
10	The sculptures: *'Labyrinthodon'*	141
11	The sculptures: *Dicynodon*	149

PART 3	A DIFFICULT... AND, PERHAPS, TOO BOLD, ATTEMPT	155
12	The reception and legacy of the Geological Court	156
13	Past becomes future: the conservation of the Geological Court	175

References	184
Image credits	189
Acknowledgements	190
Index	191

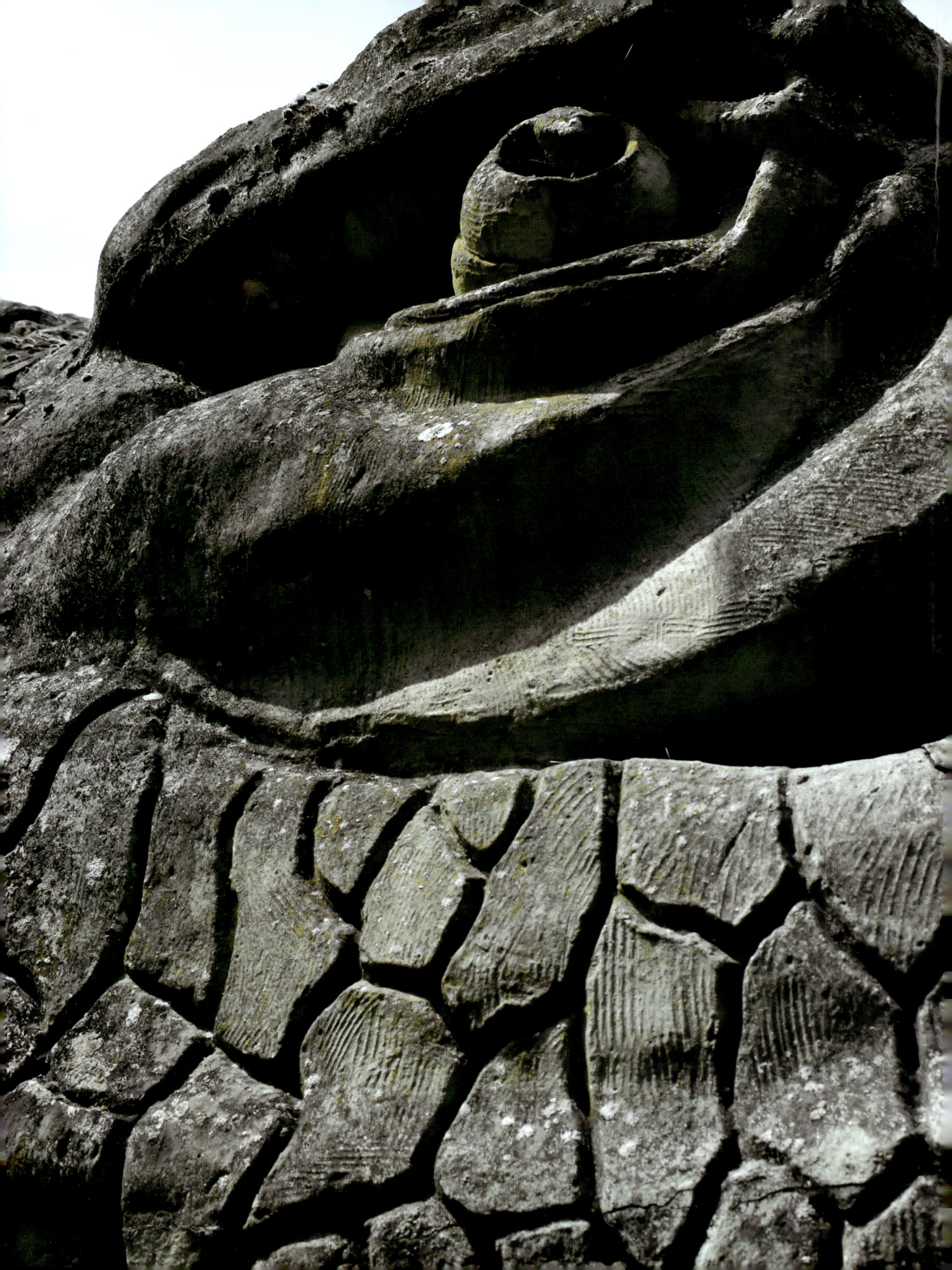

PART 1

Islands covered by strange figures

Taking our stand on the Grand Plateau, fifty feet in width, to which we have arrived, we obtain a general view of a tract of several acres of ground occupied by Geological Illustrations, and including a number of islands already partly covered by strange figures, the restored forms of various animals which for many ages have ceased to live as living tribes.

SAMUEL PHILLIPS, 1856, P. 189

The concepts of Deep Time and the existence of life on Earth before humanity are, today, such well-established and ingrained facets of common knowledge that it's difficult to imagine an existence ignorant of them. No museums full of spectacular fossils of extinct organisms. No artworks or books dedicated to the plants and animals now lost to time. And, having paused to consider the magnitude of ages represented in Earth's rock record, no personal realization of our own, almost trivial lifespans measured against the continuum of Geological Time.

This has been the reality of experience for much of our species' existence, however, and these circumstances only changed in the eighteenth and nineteenth centuries when the emerging fields of geology and palaeontology realized the significance of rocks and fossils, the study of which revealed the vast scale of our planet's long history and its changing lifeforms. By the mid-1800s these disciplines had matured to the point where some order could be applied to our understanding of the distant past. We could recognize distinct periods of time and their characteristic climates, landscapes and animals, and some of the latter were considered well enough known to seem *restorable*: that is, sufficient details of form and appearance had been deduced to permit reasonably accurate artistic recreations of long-extinct creatures and habitats without relying wholly on speculation or guesswork.

It's in this context that the story of the Crystal Palace Dinosaurs has its origins: a project by enterprising individuals, bold scientists and ambitious artists who were comfortable enough with extinct animals and geological history to attempt to capture them in concrete and iron, while also seeking to capitalize on their educational and commercial value. Our efforts to explore the artistic and scientific history of the Crystal Palace Dinosaurs and their Geological Court home must thus begin here, by appreciating the circumstances under which they were conceived and built, the people who designed and constructed them, and the philosophies that informed their execution.

CHAPTER 1

Historic prehistory in South London

The sprawling suburbs of London seem like the last place on Earth to find a long-surviving population of extinct creatures, but should you travel to Crystal Palace Park in the southeast region of the city you will encounter thirty prehistoric residents that have lived alongside Londoners for over 150 years. These antediluvian animals are, today, fairly unassuming presences on their artificial islands and many bear scars from long-term exposure to the famously changeable British weather, as well as encounters with the British public. But this inauspicious modern appearance is a pale reflection of their past glory and significance to the history of science and public education, nor does it reflect the zeal and enthusiasm they inspire among scholars, artists and the public. These are, of course, the famous Crystal Palace Dinosaurs: Victorian-age sculptures and associated landscaping that recreated extinct animals and geological features as understood by scientists in the 1850s.

The story of how dinosaurs and other extinct creatures came to dwell in suburban London is well known to anyone with a passing interest in the history of palaeontological science, and especially so among dinosaur fans. They are remnants of the Crystal Palace Park, a grand experiment with scientific communication that publicly celebrated education, innovation and commerce from 1854 to 1936. The centrepiece of this extravaganza was the titular Crystal Palace: a vast glass, iron and timber building that housed exhibitions dedicated to the very finest art, science and culture from the Victorian world. The amazing and huge exhibits held inside the Palace were only surpassed by the grandiose fountains, statues and gardens of the capacious, sloping park grounds.

Excellent records of the building and its gardens show an extraordinary commitment to scale and ambition, a feat rarely matched among purely educational and cultural projects (Beaver, 1986; Leith, 2005). But despite tremendous effort and investment, the Crystal Palace had a difficult history, and a great fire in November 1936 marked the final end of a long, arduous story. Without the Palace as a focal point, the ownership and function of the park changed significantly as its grounds were modified to suit their new role as one of London's green public spaces. Much of the park was renovated to accommodate sports facilities and concert venues, leaving only a few statues and the foundations of the Palace as reminders of long past glory days.

The Crystal Palace Dinosaurs were among these surviving elements. Tucked away in the southeast corner of the park, the prehistoric models and their manufactured landscape were untouched by the fire that consumed the Palace and they have continued to exist more or less, to the modern day entirely as they were in the Victorian era, even escaping the bombing raids that razed London in the Second World War.

Fig. 1.1 A selection of the Crystal Palace Dinosaurs, a remarkable collection of Victorian palaeontological reconstructions and recreated geological strata located in Penge, south London. Here, the three dinosaur species, *Teleosaurus*, various plesiosaurs and the Oolite outcrops can be seen – just a subset of this amazing historic site. (2015)

Fig. 1.2 The amazing Crystal Palace and Terrace at Sydenham in their nineteenth-century heyday. The Crystal Palace Dinosaurs were part of a grand exhibition of Victorian science, culture and engineering displayed in and around this enormous building. The park grounds and terrace structure can still be found in southern London today, but the Palace itself was destroyed in a huge fire in 1936.

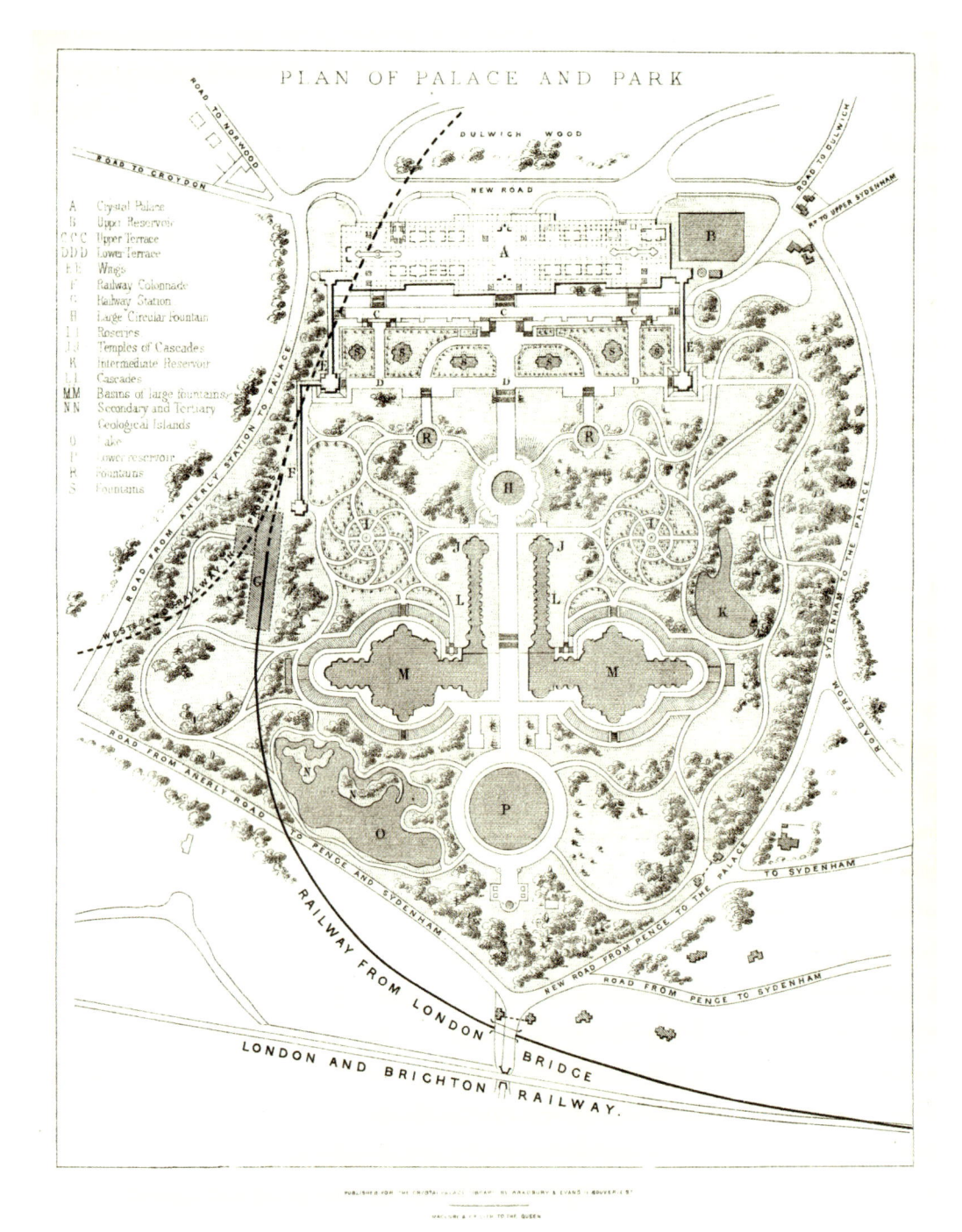

Fig. 1.3 Map of the Crystal Palace grounds, 1857. The park is approximately 1.2 × 0.8km and was, during the Crystal Palace era, filled with attractions, exhibitions, gardens and fountains. The prehistoric animals and geological landscape are located in the lower left region of the park, the geographic southeast ('N' in the legend of this map). Some details of lake size, boundaries and relative land areas are inaccurate in this graphic.

Created as representations of prehistory, the Crystal Palace Dinosaurs have since attained a historic significance of their own, becoming enduring monuments to Victorian ideals as well as witnesses to many political and cultural changes in the last 170 years.

Strictly speaking, only four of the Crystal Palace Dinosaur models are true dinosaurs, but this term has come to encapsulate the entirety of the Crystal Palace display in lieu of a more convenient or appropriate title. In truth, the label of 'Crystal Palace Dinosaurs' undersells their full extent. The entire display actually took up a whole region of Crystal Palace Park known as the 'Geological Court', which initially contained landscaped islands and lakes, numerous recreated geological features and extinct plants, as well as at least thirty-seven models of over twenty-one species of extinct mammals and mammal-relatives, dinosaurs, flying and swimming reptiles, and prehistoric 'amphibians'. Built in the early 1850s and unveiled to the public in 1854–55, many more sculptures and geological features were planned but the Court was never finished thanks to the struggling finances of

10   Chapter 1 – Historic prehistory in South London

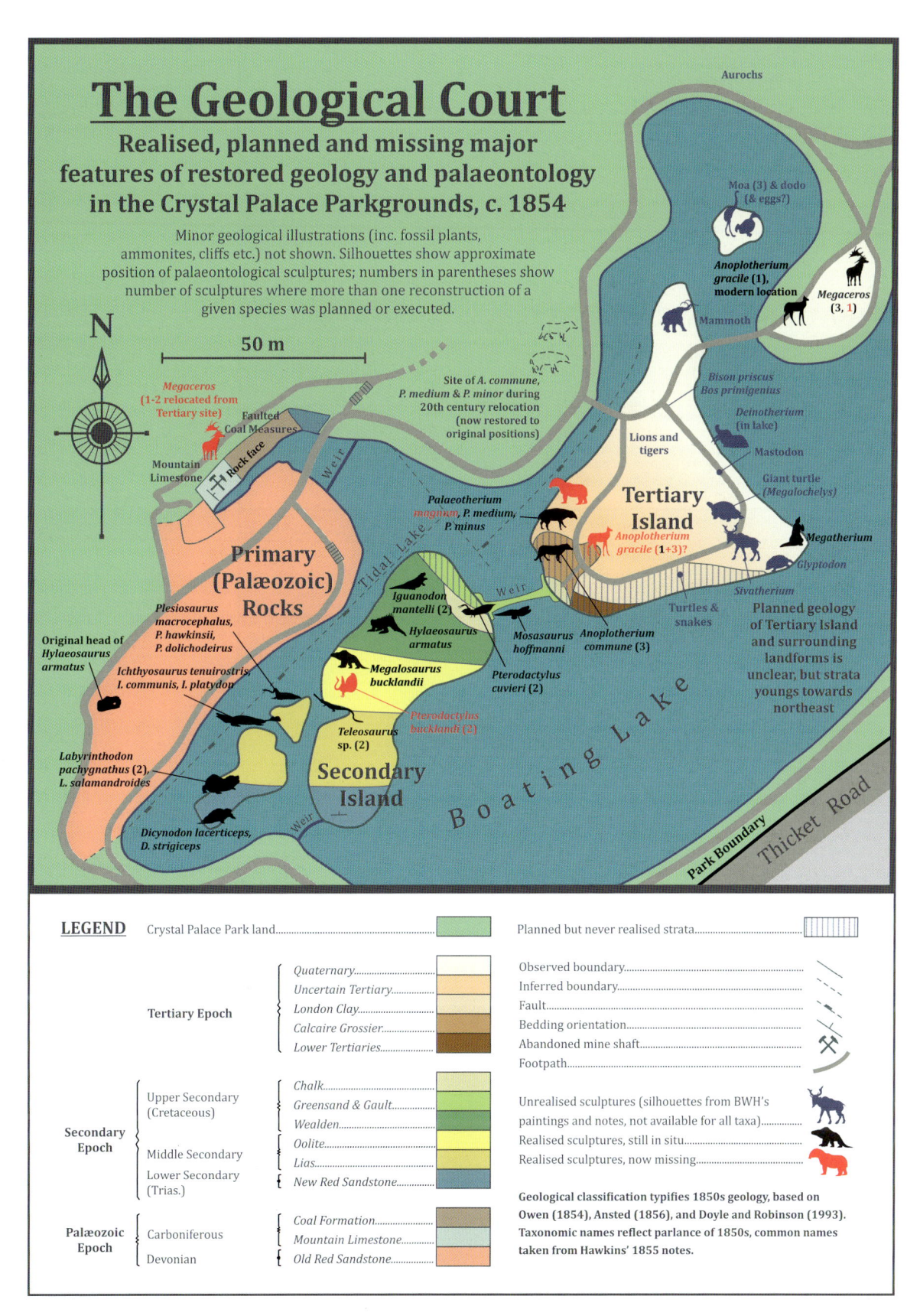

Fig. 1.4 The Geological Court, the term originally used to describe the Crystal Palace Dinosaurs and their associated geological displays, in 1854. Approximately half of the planned Geological Court was achieved before work ceased: this map shows both completed, planned and now missing components. Most of the elements built for the Geological Court survive today, albeit often in states of disrepair.

Chapter 1 – Historic prehistory in South London 11

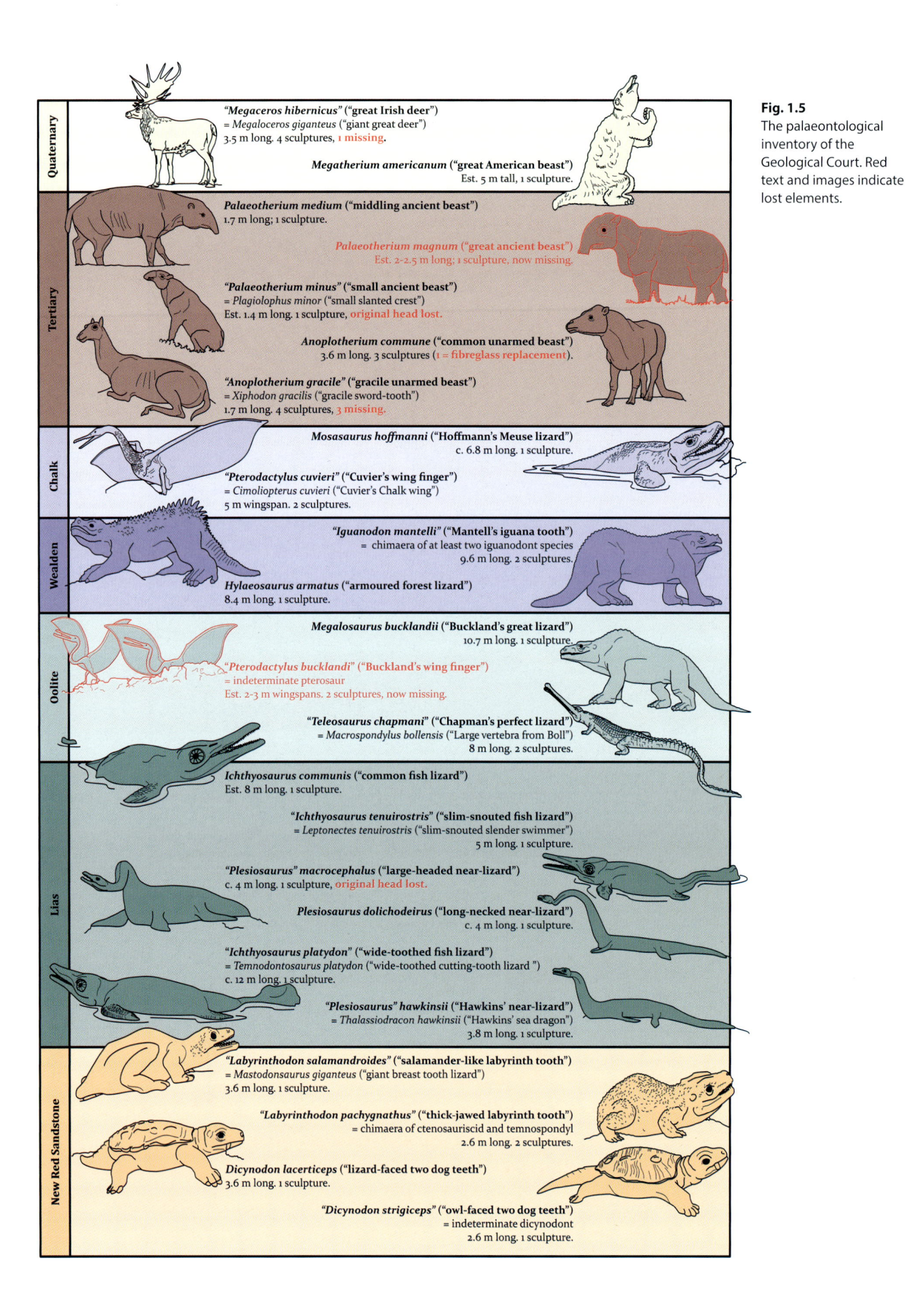

Period	Species
Quaternary	*"Megaceros hibernicus"* ("great Irish deer") = *Megaloceros giganteus* ("giant great deer") 3.5 m long. 4 sculptures, 1 missing. *Megatherium americanum* ("great American beast") Est. 5 m tall, 1 sculpture.
Tertiary	*Palaeotherium medium* ("middling ancient beast") 1.7 m long; 1 sculpture. *Palaeotherium magnum* ("great ancient beast") Est. 2-2.5 m long; 1 sculpture, now missing. *"Palaeotherium minus"* ("small ancient beast") = *Plagiolophus minor* ("small slanted crest") Est. 1.4 m long. 1 sculpture, original head lost. *Anoplotherium commune* ("common unarmed beast") 3.6 m long. 3 sculptures (1 = fibreglass replacement). *"Anoplotherium gracile"* ("gracile unarmed beast") = *Xiphodon gracilis* ("gracile sword-tooth") 1.7 m long. 4 sculptures, 3 missing.
Chalk	*Mosasaurus hoffmanni* ("Hoffmann's Meuse lizard") c. 6.8 m long. 1 sculpture. *"Pterodactylus cuvieri"* ("Cuvier's wing finger") = *Cimoliopterus cuvieri* ("Cuvier's Chalk wing") 5 m wingspan. 2 sculptures.
Wealden	*"Iguanodon mantelli"* ("Mantell's iguana tooth") = chimaera of at least two iguanodont species 9.6 m long. 2 sculptures. *Hylaeosaurus armatus* ("armoured forest lizard") 8.4 m long. 1 sculpture.
Oolite	*Megalosaurus bucklandii* ("Buckland's great lizard") 10.7 m long. 1 sculpture. *"Pterodactylus bucklandi"* ("Buckland's wing finger") = indeterminate pterosaur Est. 2-3 m wingspans. 2 sculptures, now missing. *"Teleosaurus chapmani"* ("Chapman's perfect lizard") = *Macrospondylus bollensis* ("Large vertebra from Boll") 8 m long. 2 sculptures.
Lias	*Ichthyosaurus communis* ("common fish lizard") Est. 8 m long. 1 sculpture. *"Ichthyosaurus tenuirostris"* ("slim-snouted fish lizard") = *Leptonectes tenuirostris* ("slim-snouted slender swimmer") 5 m long. 1 sculpture. *"Plesiosaurus" macrocephalus* ("large-headed near-lizard") c. 4 m long. 1 sculpture, original head lost. *Plesiosaurus dolichodeirus* ("long-necked near-lizard") c. 4 m long. 1 sculpture. *"Ichthyosaurus platydon"* ("wide-toothed fish lizard") = *Temnodontosaurus platydon* ("wide-toothed cutting-tooth lizard") c. 12 m long. 1 sculpture. *"Plesiosaurus" hawkinsii* ("Hawkins' near-lizard") = *Thalassiodracon hawkinsii* ("Hawkins' sea dragon") 3.8 m long. 1 sculpture.
New Red Sandstone	*"Labyrinthodon salamandroides"* ("salamander-like labyrinth tooth") = *Mastodonsaurus giganteus* ("giant breast tooth lizard") 3.6 m long. 1 sculpture. *"Labyrinthodon pachygnathus"* ("thick-jawed labyrinth tooth") = chimaera of ctenosauriscid and temnospondyl 2.6 m long. 2 sculptures. *Dicynodon lacerticeps* ("lizard-faced two dog teeth") 3.6 m long. 1 sculpture. *"Dicynodon strigiceps"* ("owl-faced two dog teeth") = indeterminate dicynodont 2.6 m long. 1 sculpture.

Fig. 1.5
The palaeontological inventory of the Geological Court. Red text and images indicate lost elements.

Fig. 1.6 An underappreciated aspect of the Geological Court are the Geological Illustrations: recreations of the same British rock outcrops that provided the fossil basis for the animal restorations. These simulated cliffs, escarpments and exposures were created at great expense and labour by importing the exact rock types they represent from quarries around the UK. Here, a stack of Oolitic limestone is seen behind the *Teleosaurus*. (2021)

Fig. 1.7 A northeasterly view of the Secondary Island, the home to the most famous denizens of the Geological Court. This vantage point shows a view through one hundred million years of history and would have been, for many visitors in the mid-1800s, their first exposure to concepts such as Deep Time, extinct animals and geological theory. (2015)

the Crystal Palace Project. What we see today is somewhere between one-half and two-thirds of the envisioned display.

The sculptures were primarily the work of the master natural history artist and palaeoartist Benjamin Waterhouse Hawkins and – ostensibly – his consultant Professor Richard Owen, while the geological displays and landscape were designed by geologist David Thomas Ansted and executed by the mining engineer James Campbell. Although many components of the recreated geology and animals are now scientifically dated, the Crystal Palace Dinosaurs are internationally recognized for their importance to the history of Earth sciences, the democratization of geology and palaeontology, and our capacity to visualize long-extinct organisms and landscapes. They can still be freely visited today and continue to captivate, educate and enthral the public, although some original components – including multiple animal sculptures and many of the original 'Geological Illustrations' – have been lost or damaged. Periodic rounds of conservation have made efforts of various success to reset the Geological Court to a condition like that of 1854, but the age of the recreated geological features and sculptures leaves them vulnerable to extremes of climate and reckless visitors.

The historic significance of the Crystal Palace Dinosaurs

The basic history and importance of the Crystal Palace Dinosaurs will be familiar to many readers, as will be the scientific fields they have most relevance to. They were especially critical to the development of nineteenth-century geology and palaeontology, but constant retelling of their story has seen certain details lost or mischaracterized to the extent that we must straighten out some common misconceptions before delving deeper into their history.

Among these is the idea that the Crystal Palace Dinosaurs were the first life reconstructions of prehistoric animals or, at least, the first flesh reconstructions of dinosaurs. Neither of these claims is true. The Geological Court contains a few reconstructions that, to our knowledge, might be the first attempts at restoring the life appearance of certain species (*Mosasaurus*, the giant pterosaurs, *Dicynodon* and others) but most of the extinct animals at Crystal Palace had been restored by other artists before 1854 – albeit sometimes in dramatically different guises. Dinosaurs were among these, being first recreated in art during the 1830s.

Fig. 1.8 The Crystal Palace Dinosaurs were not the first life reconstructions of extinct animals but actually followed several decades of palaeoartistry. They are, however, highly significant works of early palaeoart that equal other important early works of the genre, including **(A)** Henry De La Beche's *Duria Antiquior* (1830), the first fully realized palaeoartwork, showing a scene of Jurassic Dorset; and **(B)** Georg August Goldfuss's *Jura Formation* (1831), an illustration based on and advancing some of the concepts shown in *Duria Antiquior*. Such artworks represented the conceptual acme of palaeoart until the development of the Crystal Palace Dinosaurs in the early 1850s.

Fig. 1.9 Early to mid-nineteenth-century palaeoart of fossil reptiles depicted radically different, and substantially less accurate, reconstructions to those featured at Crystal Palace. **(A)** *Reptiles Restored, the Remains of Which Are To Be Found in a Fossil State in Tilgate Forest, Sussex* (1833) by George Scharf, featuring *Iguanodon*, *Hylaeosaurus* and other reptiles; **(B)** *The Ancient Weald of Sussex* (1838) by George Nibbs, with *Iguanodon* and *Megalosaurus* alongside marine reptiles; **(C)** John Martin's *The Sea Dragons as They Lived* (1840); **(D)** Martin's *The Age of Reptiles* (1842), showing plesiosaurs, an assumed ichthyosaur and pterosaur – the identity of the largest reptile is uncertain – it is possibly *Megalosaurus*, based on its terrestrial habits and similarity to Martin's other dinosaur artwork; **(E)** frontispiece of John Mill's *The Fossil Spirit: A Boy's Dream of Geology* (1854), featuring an assumed *Iguanodon*; **(F)** Josef Kuwasseg's *c.* 1850 *Iguanodon* from Franz Unger's *Die Urwelt in ihren verschiedenen Bildungsperioden*, one of the most influential dinosaur reconstructions of the nineteenth century.

Indeed, palaeoart, the specialized branch of natural history art concerned with the reconstruction of extinct organisms and ancient landscapes, was already a well-established genre before work commenced on the Geological Court. The concept of scientifically reconstructing extinct organisms in art seems to have begun around 1800 (Taquet and Padian, 2004), so the Crystal Palace Company's idea of restoring extinct species was far from novel. What *was* novel, however, was the creation of life-size, three-dimensional reconstructions: before Crystal Palace, most or all palaeoart was two-dimensional and produced to scale.

Another misconception is that the Crystal Palace Dinosaurs represent the earliest ideas of how scientists interpreted extinct life. This is also untrue. Our knowledge of fossil organisms was, in some cases, decades ahead of that which informed the first life restorations of extinct animals, such that the sculptures at Crystal Palace actually represent second or third generations of palaeontological interpretation. This is especially so for the three dinosaur species which, in being more obviously 'mammalian' than stereotypically 'reptilian' in form, had advanced from notions of dinosaurs as giant, lizard-like animals towards modern concepts of dinosaurs as upright, active, warm-blooded creatures. It's accurate to regard the Geological Court as capturing *early* interpretations of extinct life, but they represent several decades of scientific advancement from the first.

A third major misinterpretation of the Geological Court restorations is that they are, by virtue of their age, archaic and simplistic restorations that are not as sophisticated as modern palaeoart. While the vintage nature of the science informing the Crystal Palace sculptures is undeniable, their rendering and execution is far from simplistic. To the contrary, the Crystal Palace Dinosaurs have an attention to anatomical detail and form that transcends many later palaeoart sculptures, including many made today. In terms of presenting realistic posing and behavioural depictions, the palaeoart of Crystal Palace was far ahead of the fantastical, melodramatic compositions which had characterized the genre in the early 1800s. Many early artworks of fossil reptiles, for example, owe much to classical depictions of dragons with their serpentine spiralled tails and gurning, toothy mouths (*see* Fig. 1.9). For all their scientific flaws, Hawkins' extinct reptiles avoided all these conventions and look conceivably real.

These common misconceptions demonstrate that appreciating the Crystal Palace Dinosaurs is reliant on an accurate

Fig. 1.10 Often thought of as an especially early event in palaeontological history, the Geological Court was actually built upon several decades of geological and palaeontological work. Fossils of species like *Megaloceros* had been known for over 150 years before the Crystal Palace Park was conceived. (2017)

understanding of their historic context. Geological disciplines had been studied for several generations by the time the Crystal Palace Dinosaurs were constructed, such that, while our knowledge of extinct organisms, Deep Time, and the geological evolution of our planet was still very incomplete, we were far from naive or clueless about such topics. Geologists and palaeontologists had been constructing scientifically-based, evidence-led interpretations of prehistoric Earth since the 1700s, and developments in geosciences moved rapidly throughout the nineteenth century. Fuelled by sharpened scientific principles and revelations about the nature of reality discovered during the seventeenth and eighteenth centuries (the 'Age of Enlightenment'), early Earth scientists documented vast amounts of geological and palaeontological data and established fundamental concepts essential to the study of prehistoric Earth. These included the extreme age of our planet, the changing nature of our landscapes and lifeforms, and biological extinction. More sophisticated understandings of these topics and other important theories would come later, including biological evolution through the process of natural selection, the accruement of vast fossil collections representing the diversity of life through time, and an accurate estimate of the age of the Earth. But the work of geologists and palaeontologists in the early nineteenth century laid the groundwork for a radical re-thinking of humanity's place in history. Life on Earth had existed before humankind, perhaps for a considerable amount of time, populated by now-extinct lifeforms unrecognizable to modern eyes.

But while the academic world was increasingly comfortable with reconstructing the past through scientific means, the general public of the early 1800s remained mostly unaware of prehistoric worlds and long-extinct animals. Museums existed but had yet to develop the extensive natural history collections, universal public access, and accessible interpretation methods that we associate with museums today. Indeed, the very concept of our modern natural history museums was a development of the late nineteenth century: a product of growing interest in biology and geology and the realization that such disciplines warranted their own archives and public galleries (Farber, 1982). The public displays of geology that existed before this were sombre, cabinet-lined rooms full of catalogued but largely unexplained extinct shellfish, broken fossil bones and geological specimens, and thus not especially welcoming or accessible to lay audiences. These would have been exciting treasure troves for specialists, but a lack of interpretation and explanation would have made them uninformative and uninteresting to the public at large.

On top of this, a major element missing from geoscience communication in the early 1800s was palaeoart. The first palaeoartworks (*see* Fig. 1.8) were produced by the same learned figures that were studying rocks and fossils for their own scholarly pursuits, and their artworks were almost exclusively circulated among wealthy academics (Rudwick, 1992; Lescaze, 2018; Witton, 2018). The strength of palaeoart as a public education tool – with its ability to communicate much about prehistoric life without the need for lengthy explanations or scientific jargon – had not yet been identified, and use of the genre only stepped slightly in this direction when palaeoart entered use in universities as a teaching aid. Geology and palaeontology were popular hobbies in the nineteenth century but palaeoart was not yet a major facet of these interests, and it was never given a grand stage when featured in books or periodicals. Early scholars saw the reconstruction of fossil animals as a useful illustrative exercise, but also as secondary to primary data sources such as fossils or restored skeletons.

The Crystal Palace Dinosaurs represented a marked departure from this. This grand public display featured cutting-edge geological and palaeontological science in an accessible and exciting fashion, showcasing new and wholly unusual fossil animals discovered in the early 1800s alongside a geological backdrop that emphasized the antiquity of the Earth. For visitors without the time and resources to pursue geology as a hobby, this was surely the first time they would have come face-to-face with these mind-bending concepts.

Thanks to the reality of life before humanity, spectacular extinct species, and the unfathomable depth of geological time being rote parts of most childhood educations today, we can only imagine what it was like to learn these facts as older, more experienced individuals who had previously felt confident in their place in the world. What must it have been like to learn that giant reptiles once walked where we now stand, or that the rocks and landscapes around us were far older than anyone had imagined? The construction of the Geological Court allowed core concepts of Earth sciences to enter public consciousness in a major way: an important, spectacular and creative introduction of prehistoric life and geological principles to mainstream society.

It is no exaggeration to say that the Crystal Palace Dinosaurs were a great influence on the next two centuries of palaeontological outreach and artwork, as well as the commercialization of prehistory. Though educational in purpose, the Crystal Palace was a private enterprise that monetized enlightenment: if you wanted to see the best of Victorian culture, engineering and science, you had to purchase a ticket. The Geological Court can thus be viewed as a rare fusion between a major business opportunity, palaeontological science and public education, leading to an impressive list of accolades. This was the first public-focused exhibit of life before humanity; the first attempt to reconstruct models of ancient species and landscapes at life-size; the first investment of large amounts of capital into the portrayal of prehistoric life; the first realization of the formidable power of prehistoric subjects in advertising and merchandising; and the first palaeoart project to explicitly emphasize the fusion between science and artistry.

It is little wonder that, for several decades after their construction, many artists simply replicated the Crystal Palace restorations when executing artwork of prehistoric animals. It is difficult to think of a singular event combining the same corporate, palaeontological and artistic interests at such scale, and having the same cultural impact, as the Crystal Palace Dinosaurs. Perhaps the release of the 1993 film *Jurassic Park* is the only comparable milestone: a film which famously modernized portrayals of dinosaurs using state of the art special effects and cutting-edge palaeontological science, and had a similar game-changing effect on public concepts of prehistoric life.

Today, the Crystal Palace Dinosaurs have a devout following among palaeontological enthusiasts. They are seen as a window into mid-Victorian geosciences: a physical reminder of the ideas, philosophies and individuals who shaped the early development of palaeontology and geology. They have

Fig. 1.11 Several displays in the Geological Court have become truly iconic, highly influential works of palaeoart. Among them are the two *Iguanodon* reconstructions, famously restored as rhinoceros-like quadrupeds owing to our partial understanding of their anatomy in the 1850s. These models are often characterized as wholly inaccurate, but actually represented a significant step forward in our reconstructions of dinosaurs. (2017)

become part of the established canon of palaeontological history with countless books, articles and documentaries recounting their significance to the study of extinct life, and especially the early study of dinosaurs.

Plans to produce prehistoric statues during the mid-nineteenth century were hatched elsewhere, such as France (Knoll and López-Antoñanzas, 2010), the United States (e.g. Bramwell and Peck, 2008) and Russia (*see* Chapter 12), but most were never executed, making Crystal Palace a unique and important insight into nineteenth-century palaeoart and science communication. The Geological Court has accordingly been the focus of much academic interest, inspired a number of artists and authors, and attracts visitors from distant parts of the globe to see the only life-sized Victorian models of extinct animals on the planet.

Not all coverage of the Crystal Palace Dinosaurs has been positive, however, and they have long endured detraction from critics about their outdated depictions of prehistoric life. The cruellest criticisms see them cast as laughable or misleading attempts at reconstructing fossil worlds and, for being so outdated, having no relevance to modern science. Remarks of this nature are not new, being shared by critics from the day the park opened in 1854, and are still made today.

To anyone interested in the history of palaeontology and palaeoart, some of these observations are obvious to a degree of pointlessness, and they also miss the point that all

Fig. 1.12 One of the best-known images of the Crystal Palace Dinosaurs: an engraving of Benjamin Waterhouse Hawkins' workshed based on a photograph taken by Philip Henry Delamotte. This image was created for promotional purposes, appearing in a December 1853 edition of the *Illustrated London News*: even in the 1850s, the star-power of extinct animals was a powerful advertising agent. The clay mould of the *Iguanodon* dominates the scene, surrounded by *Palaeotherium magnum*, *Hylaeosaurus*, '*Labyrinthodon pachygnathus*' and '*Dicynodon strigiceps*'.

artistic and scientific work must be evaluated in an appropriate historic context. *Of course* the Crystal Palace Dinosaurs are scientifically outdated, as they were by the late 1800s, because they represent an early interpretation of extinct life in a fast-moving scientific discipline now over two centuries old. A fairer and more nuanced take on the Geological Court and its inhabitants views them as a fascinating case study in scientific communication. The relative infancy of Earth sciences in the 1850s may have made it presumptuous or arrogant to assume we could capture in concrete and iron what had only been partially understood from fossils but, conversely, the history of science shows that few hypotheses and theories escape revision over time, and that science communicators can do no more than present snapshots of contemporary understanding to the public.

There is much to discuss around this point, both in context of vintage examples like the Crystal Palace Dinosaurs as well as the modern sensationalization of scientific stories that prioritize publicity over academic rigour. To disregard the Crystal Palace Dinosaurs for merely being old and scientifically inaccurate is not only unfair to their creators, but also ignores the lessons they impart about the scientific process and public education.

A familiar story, an enduring enigma, and an ongoing conservation risk

It is this sort of discussion that underscores the iconic status of the Crystal Palace Dinosaurs. The core elements of their development are not only mentioned in innumerable books for readers of all ages, but are the focus of several excellent academic texts, papers and book chapters (e.g. Rudwick, 1992; Doyle and Robinson, 1993; McCarthy and Gilbert, 1994; Secord, 2004; Doyle, 2008; Bramwell and Peck, 2008). A large number of historic documents – personal correspondence, newspaper and magazine articles, photographs and illustrations – record the development and critical response of the Geological Court, as well as the politics and personalities involved in their creation. From these, a reasonably detailed picture of the objectives and aims of the Geological Court, how it came to be, and its historic legacy have been established.

And yet, despite these records, many points about the Geological Court remain debated or mysterious. Few documents specify the process of construction or the scientific rationale behind the displays, and some records offer confused or conflicting information. Some of these mysteries are minor, such as uncertainties about the specific logistical arrangements for the famous 1853 New Year's Eve banquet in the clay *Iguanodon* mould. Others are major, such as how many sculptures were originally built for the park. Twenty-nine originals and one accurate replacement are *in situ* today, but historic records show that at least another seven – two pterosaurs and five mammals – incontrovertibly once existed: a total of thirty-seven (*see* Figs 1.4 and 1.5).

Another concern is the lack of documentation concerning the damage and repairs made around the Geological Court in the last 170 years, including those related to park redevelopments that removed or destroyed large parts of the display. Such issues are not historic minutiae or simply 'nice to know' information: they guide and shape the ongoing conservation efforts at the park. The result is that a surprising amount of detective work is essential to unravelling the story of the Geological Court, and new resources and connections between data points are still coming to light.

It is in this spirit that the following book has been written. One investigative approach that has hitherto been neglected is to examine the Geological Court as the world's first major palaeoart project, and in the process uncover the science, theory and ideas captured in these historic Geological Illustrations and palaeontological sculptures. Palaeoart may outwardly seem to be a simple artform – find the fossils of an extinct organism, arrange them in a life-like pose and sketch in the missing parts – but it is actually a highly involved, complex means of researching and representing scientific hypotheses about the appearances of extinct organisms and landscapes in art (Witton, 2018). A finished palaeoartwork is a snapshot of scientific and artistic thinking that reflects numerous influences, including the palaeontological and geological data available to the artist; the interpretations of extinct organisms and their worlds at the time of execution; the assumptions and speculations made to plug gaps in our understanding of the subject species and habitat; and the artist's intention and stylistic choices. Examining the Crystal Palace Dinosaurs as works of palaeoart permits insights into Victorian concepts of prehistory that are not captured in other historical investigations, shines new light on their construction, and provides further insight into the enduring mysteries of the Geological Court. What does the muscle distribution of *Megalosaurus* tell us about early ideas of dinosaur lifestyles? Why did the (now missing) *Palaeotherium magnum* sculpture have such elephantine features despite this species being related to horses? Why do some details of the sculptures directly contradict the data published by their primary consultant? Such questions provide a fresh perspective on these oft-discussed models while also helping us to appreciate their artistic sophistication. Although scientifically dated, there is nothing simplistic or primitive about the execution of the Crystal Palace Dinosaurs themselves.

This attempt to re-examine the Geological Court coincides with a wider push for recognition of the Crystal Palace Dinosaurs as worthy and significant parts of not only scientific history, but British heritage in general. The sculptures and their landscape represent geological public engagement of a nature and vintage unseen anywhere else in the world, as well as representing the last largely untouched remnant of the original Crystal Palace Park project. They are priceless, irreplaceable Victoriana that deserve the same level of care and preservation as the most carefully curated museum specimens.

Alas, the very properties that make the Geological Court special are also the biggest challenges to its long-term conservation. The entire site is exposed to the elements and the displays are vulnerable to trespassers who, by intent and accident, routinely damage the models and their surroundings. The history of the Court is a constant battle against deterioration and quests for conservation funding. Since the 1950s, conservation of the Geological Court has happened periodically but this has not stopped some models deteriorating to an incredible extent. Many are far more reconstructed than they outwardly appear and a number of original components have vanished completely from the park, sometimes directly due to errors in conservation efforts. The need for repair and maintenance is continuous, but only some demands are met in time to avoid chronic escalation of damage and decay.

This is not to say that the Crystal Palace Dinosaurs are neglected or ignored by relevant authorities. In 1973 the sculptures were classified as Grade II Listed Buildings on Historic England's National Register of Heritage Monuments, and are thus protected from modification without special permission. They were upgraded to Grade I, the UK's highest level of heritage recognition, in 2007 and, in 2020, were placed on the Historic England's 'Heritage At Risk' register – the highest priority for conservation.

The continued plight of the Geological Court has seen concerned citizens taking increasing interest in its conservation and maintenance, too. In 2013 the 'Friends of Crystal Palace Dinosaurs', a registered charity dedicated to the conservation and maintenance of the Geological Court, was formed by local residents and academics. This group, which has grown to encompass members and project partners across the UK, provides advice and assistance to the legal custodians of Crystal Palace, London Borough of Bromley, to preserve, document and maintain the site, research its history, and engage in outreach exercises to better communicate its importance to the public. It also spearheads funding initiatives to complete conservation-related projects, the most ambitious of which was the building of a permanent, securable bridge across the waters surrounding the Secondary Island (the landmass featuring dinosaurs and other fossil reptiles) to give easier access for conservation purposes. This was successfully funded in 2019, with the bridge installed in 2021.

These conservation achievements are victories in the long-term battle to conserve the world's oldest geological park and its life-sized models of prehistoric species, but we must maintain a realistic view of their long-term prospects: the Crystal Palace Dinosaurs face an uncertain future unless their management and conservation become more regular and reliable. It is in response to this, and especially damage incurred to the *Megalosaurus* sculpture in May 2020, that catalysed this book. Suspected human interference saw the front of the *Megalosaurus* face break off along pre-existing cracks and weaknesses, exposing the internal armature, bricks and other historic building materials in a manner that was difficult not to compare with a badly wounded animal. The *Megalosaurus* has since been fitted with a prosthetic jaw, but its injured face offered a clear vision of what will happen to the entire Geological Court if maintenance and conservation are not stepped up. The struggles of this important site against deterioration need to be a matter of wider knowledge and concern.

But this is not our only objective: we also want to provide the most detailed insight into the Geological Court and its prehistoric denizens yet compiled. We are both publishing new data based on our own findings as well as synthesizing accounts by previous researchers and archivists to weave as much information as possible into our narrative. We have received a lot of gracious help to put this book together but, even so, our story remains incomplete: there are many questions which we are currently unable to answer. What you're about to read is our understanding of the Geological Court in 2021, and we hope that our work will enthuse others into learning more about the Crystal Palace Dinosaurs and their park home. We specifically hope that our work will excite *you*, our readers, into having not only a greater interest in the Geological Court, but more reverence for the work that went into it and its place in history.

As for our third and final goal, we are pleased to say that we have already achieved it. By buying this book, you have already made a monetary contribution to funding the Friends of Crystal Palace Dinosaurs' work in the Geological Court – thank you. But our interaction does not need to stop here: the Friends offer occasional tours of the Geological Court, volunteer positions for outreach and site maintenance, and other ways to get involved with Victorian dinosaur conservation. Please consider turning this introduction into a longer relationship by following and supporting their work, and visit cpdinosaurs.org.uk to find out how you can directly help their efforts.

Fig. 1.13 The imposing face of the Crystal Palace *Megalosaurus*, as seen in 2021, sporting a prosthetic lower jaw after recent conservation work. Protecting the Geological Court from a multitude of conservation risks, including extremes of weather, aggressive plant growth and human vandals, is a constant need, and many features of the site have suffered badly from these agents.

CHAPTER 2

Ancient worlds through a Victorian lens: planning the Geological Court

Even for a culture as enamoured with spectacle and self-aggrandizing as the Victorians, the development of a twenty-acre site celebrating geology and palaeontology was an unusual exercise. Nothing of the kind had been attempted before and efforts to visualize the past through artistry were restricted in number and scale. Where, then, did the idea of building life-sized models of extinct animals and simulated rock outcrops come from?

At some fundamental level, the Geological Court can be seen as an especially eccentric example of Victorian Britain's enthusiasm for large, ostentatious engineering and scientific projects. The mid-1800s can be viewed, from a nationalistic perspective, as Britain's 'golden years': a time of great industrial and military might, relative prosperity and empirical expansion. The Victorians invested heavily in ambitious and expensive projects around the UK, including grand buildings and monuments, a national railway and road system, a network of sewers and underground transit system for London, as well as numerous innovative bridges and tunnels.

Science and research boomed among the mid-nineteenth-century Britons, heralding the invention of photography, revolutionary communication technologies such as telephones and telegraphs, numerous advances in medicines, and groundbreaking developments in geology, astronomy and natural history. For those enthused about knowledge and enterprise, it's difficult not to view this period in British history as an exciting, romantic era of discovery and innovation, though we must remember that Victorian Britain was seen very differently by the people and nations they occupied and exploited. Victorian decadence, including the celebration of advancing technology and learning that we still benefit from, had a substantial human cost.

The seed that grew into the Crystal Palace Dinosaurs was sown a few years before their conceptualization was fleshed out in detail. It was the celebration of Victorian zeal for industry, engineering and science at the 'Great Exhibition of the Works of Industry of All Nations' which originated the concept of the Crystal Palace Park and, by extension, its Geological Court. The Great Exhibition ran from May to October 1851 under the roof of an enormous temporary glass and iron building – the 'Crystal Palace' – in London's Hyde Park. Designed by Sir Joseph Paxton (1803–65), the Crystal Palace was a vast, three-storey structure over 500m long and large enough to house tall trees, grand sculptures and industrial machinery. It showcased the best of culture and technology from around the world (although obviously emphasizing content related to the British Empire) with exhibits dedicated to science, industry, exotic artefacts, art, new inventions, and music.

The Great Exhibition was a tremendous success against all popular, financial and academic criteria, such that its closing

Fig. 2.1 Before Crystal Palace Park, there was the Crystal Palace of the 1851 Great Exhibition of the Works of Industry of All Nations, more generally referred to as 'The Great Exhibition'. This temporary Crystal Palace in Hyde Park, London, housed exhibits devoted to world-leading engineering, science and art, and its popularity directly led to the development of the permanent Crystal Palace Park. Image from Joseph Nash et al. (1852), *Dickinson's Comprehensive Pictures of the Great Exhibition of 1851*.

Fig. 2.2 **A)** The creator and architect of the Crystal Palace and its gardens, Joseph Paxton (1803–65); **B)** geologist and designer of the Geological Court, David Thomas Ansted (1814–80).

22   Chapter 2 – Ancient worlds through a Victorian lens: planning the Geological Court

Fig. 2.3 The extensive gardens of the Crystal Palace Park as shown by J. Needham's 1854 lithograph, *The Crystal Palace and Park*. The skyscraping capability of the fountains are exaggerated here, but the vast layout of the park is accurate. Note the imagined complete Geological Court menagerie in the foreground.

in October 1851 was lamented and options were considered to prolong its existence. Public opinion was for the glass building to remain in Hyde Park, but existing agreements forbade it from permanently occupying this space. A provisional agreement was struck for the Palace to remain in place until 1852 while parliament decided its fate, but pressure from politicians opposed to the progressive ideals encapsulated by the Palace saw it dismantled while its future remained uncertain. This decision was not entirely a loss, however, as surplus money raised by the exhibition was poured into further funding of science, industry and education. This included the purchasing of land in South Kensington for the construction of academic institutions and collections, including the sites where the Science Museum, Victoria and Albert Museum, Natural History Museum, Imperial College, Royal College of Art and the Royal College of Music now stand.

Despite this setback, enthusiasm for the Great Exhibition was so great that even dismantling the Crystal Palace itself could not dissuade interested parties from securing it a new, permanent home. As the glass panes and iron girders were deconstructed, a consortium of businessmen, including Paxton, raised funds to relocate and rebuild the Palace elsewhere in London. This group represented the founders of the Crystal Palace Company, the owners of the Palace and grounds until their bankruptcy in 1909. They identified a suitable 300-acre location in Penge for a new, even larger Palace and commenced building their vision on the summit of Sydenham Hill in 1852. This new building used components of the original Crystal Palace but was essentially a new and much larger structure (*see* Fig. 1.2). At five storeys tall and now sporting a vaulted roof, the floor space exceeded that of the original by over 50 per cent.

The Penge site included extensive grounds on the southeastern slope of Sydenham Hill, which were developed with equivalent flair and extravagance. Among picturesque terraces, mazes, statues and gardens, fountains were constructed of such size that they required enormous reservoirs and water towers to supply the huge volumes of liquid they could thrust into the air. Said to be the biggest in Europe, the running costs of these fountains were enormous, as were the expenses of the wider Crystal Palace project. The Great Exhibition had cost £150,000 (approximately £17 million, when adjusted for inflation) but the Penge redevelopment cost almost nine times as much: £1,300,000 (over £133 million today).

Fig. 2.4 Historic courts of the Crystal Palace, as depicted in Matthew Digby Wyatt's *Views of the Crystal Palace and Park, Sydenham* (1854). The Geological Court was a natural extension of the other exhibits created for the park, where historic and ancient architecture were recreated in exacting detail by sculptors and consulting specialists. Shown here are the Renaissance, Assyrian and Egyptian Courts – just a fraction of the many spectacular exhibits built within the Palace.

As with its Hyde Park predecessor, the new Crystal Palace was filled with exhibitions celebrating culture and technology. The *Guide to the Crystal Palace and Park* by Samuel Phillips (1854) – the first of many iterations of this guide – gives a detailed and insightful glimpse of the philosophy behind the chosen content, as well as the lofty goals of the Crystal Palace Company. The stated intent was the immersion of visitors in intellectual, cultured pursuits or, as Phillips put it:

> To raise the enjoyment and amusements of the English people… in wholesome country air, amidst the beauties of nature, the elevating treasures of art, and the instructive marvels of science, an accessible and inexpensive substitute for the injurious and debasing amusements of a crowded metropolis:– to blend for them instruction with pleasure, to educate them by the eye, to quicken and purify their taste by the habit of recognising the beautiful – to place them amidst the trees, flowers and plants of all countries and all climates, and to attract them to the study of the natural sciences, by displaying their most interesting examples – and making known all the achievements of modern industry, and the marvels of mechanical manufactures:– such were some of the original intentions of the first promoters of this National undertaking.
>
> PHILLIPS, 1854, P. 13

The significance of what Phillips (1854) summarized elsewhere as a 'high moral and social tone' (p. 16) in relation to the great cost of the Crystal Palace enterprise is not to be overlooked. The Crystal Palace Company felt that education and intellectualism would be enough of a public draw to recoup their vast financial outlay and eventually turn a profit. This ideology seems striking against our modern age where business enterprises readily spend hundreds of millions on sports and entertainment, but rarely on public education.

Among the exhibits were 'courts' showcasing historic art from ancient Egypt to the modern day. As was the style at the time, these exhibits were no half measures: they included restorations of enormous Egyptian statues, replicated Grecian courts and casts of historic artworks. Visitors thus had the experience of seeing these geographically and temporally distant locations not as ruins or artefacts, but reconstructed as they would have looked hundreds or thousands of years ago. The specific use of *recreated* historic content, rather than real (or replica) artefacts, is noteworthy, as it distinguished the Crystal Palace educational experience from that of a museum. Visitors were not left to form their own conclusions from a display of genuine, unmodified historic specimens, but would instead see the past revived and restored in line with the contemporary interpretations of Victorian scholars.

The natural world was similarly recreated with plants and animal taxidermy displayed alongside one another in geographic context with details of local peoples and races. Visitors would have noticed the scarcity of display cases and

Chapter 2 – Ancient worlds through a Victorian lens: planning the Geological Court signage, a deliberate effort to 'prevent the monotony that attaches to a mere museum arrangement, in which glass cases are ordinarily the most prominent features' (Phillips, 1854, p. 16). More cynically, this lack of interpretation also meant visitors had to purchase guidebooks, of which seventeen were available.

It is easy to see how these principles were extended to create an exhibit celebrating pre-human life – the Geological Court. It is not clear who originated the idea of a geological display with prehistoric animal sculptures, but many founding members of the Crystal Palace Company had backgrounds in industrial geology and it was probably not a large intellectual leap for these individuals to extend their portrayals of human history to pre-human times (Doyle, 2008). The geologist Peter Doyle, who has significant expertise in the Geological Court, suggests that Paxton, Richard Owen or Prince Albert were likely sources, with Paxton as his preferred candidate (Doyle, 2008). Whoever came up with the idea, the Geological Court would be the first major display of its kind anywhere in the world and it would embody the same restorative principles as the rest of the Palace exhibits: a physical recreation of how leading Victorian geologists and palaeontologists interpreted life of the past, unblemished by display cabinets, broken, dusty fossil specimens or even interpretative signage.

The geological and palaeontological display envisaged by the Crystal Palace Company was, like the rest of their endeavour, an enormous project that would not be easily or cheaply achieved. Displaying prehistory to the public would require not only an ability to wrangle with cutting-edge, contemporary science, but also solve major structural and engineering problems: how would one approach the construction of a twenty-acre artificial geological landscape containing several gigantic, multi-tonne dinosaur models? Who had the sufficient scientific and artistic skills to take on these unprecedented roles? The Crystal Palace Company had its work cut out to find the right individuals to realize their ground-breaking display.

Architects of prehistory

To achieve their ambitious geological and palaeontological wonders, the Crystal Palace Company created roles within their Natural History Department specifically for the management and construction of the Geological Court (Craddock, 2016). We can be certain that whole teams of people were involved in constructing the display, but the full roster of contributing individuals has been lost over time, leaving only the project leads and consultants known to us.

Professor David Thomas Ansted (1814–80), Director of Physical Geography, Geology and Mining

The principal designer of the Geological Court, and thus perhaps the man who can be viewed as its grand architect, was Professor David Thomas Ansted (*see* Fig. 2.2B). Ansted was a geologist who held eminent titles with professional bodies and had authored numerous books on geological topics. He held the title of 'Director of Physical Geography, Geology and Mining' within the Crystal Palace Company and was a Professor of Geology at Kings College, London in

Fig. 2.5 The famous 'Baxter Print' of the Crystal Palace grounds, likely produced in early 1854 by London printer George Baxter, is one of the earliest known visualizations of the Geological Court. It uses forced perspective to make the display much larger than it was in relation to the Palace, and its creation before the park was finished meant Baxter had to follow Geological Court plans and his own imagination to complete the illustration. The reptile sculptures may be based on the Wisbech Museum models (*see* Chapter 3) and their layout – along with the Geological Illustrations – only approximates their real condition.

Fig. 2.6 The creators of and influences on the palaeontological reconstructions of the Geological Court. **A)** artist and sculptor Benjamin Waterhouse Hawkins (1807–94); **B)** palaeontological consultant Richard Owen (1804–92); **C)** Gideon Mantell (1790–1852); **D)** George Cuvier (1769–1832).

the early 1850s, before moving to the College of Civil Engineers, Putney, later in his career.

Ansted worked with Paxton to design the layout of the Geological Court while the grounds of Crystal Palace Park were being planned, placing the geological displays in an arrangement accurate to the understanding of British geology at the time. Ansted's original designs for the Geological Court are lost (Doyle, 2008), but they reflect the broad picture of geology outlined in his 1856 book *Geological science: including the practice of geology and the elements of physical geography*. This book contained several recommendations for readers to visit the grounds of the Geological Court to fully appreciate the palaeontological sculptures created by Hawkins although, perhaps modestly, they were not directed to view Ansted's equally impressive geological creations.

James Campbell (birth and death years unknown), Assistant Engineer

Although seemingly key to the creation of the geological features of the Court, geologist James Campbell is among those barely known individuals who worked on the Crystal Palace project. Campbell was an experienced mining engineer (McDermott, 1854; Doyle and Robinson, 1993; Doyle, 2008), who held the role of board member and Assistant Engineer within the Crystal Palace Company (Craddock, 2016). He was responsible for the physical execution of the geological plan created by Ansted, including sourcing various stones and rocks from around Britain needed to create the displays (Phillips, 1856). As discussed in the next chapter, the complexity and detail of the Crystal Palace Geological Illustrations is commendable, and it is most unfortunate that we do not know more about Campbell's role in shaping the landscape of the Geological Court.

Benjamin Waterhouse Hawkins (1807–94), Constructive Artist for the Restoration of Extinct Animals

Often considered of secondary importance to the Crystal Palace Dinosaurs against the historic weight of his consultant, Richard Owen, Benjamin Waterhouse Hawkins (*see* Fig. 2.6A) is probably the most significant person in the story of the Geological Court and the reason it has an enduring legacy. Hawkins was the designer and sculptor of the palaeontological sculptures and it is his work – especially in light of recent revelations about Owen's absent consultancy (see below) – that has proven to be the most famous and celebrated component of the Geological Court.

Hawkins, who preferred to go by his late mother's maiden name, Waterhouse, over Benjamin, has long been a somewhat mysterious historic figure, but his life and work have been pieced together by his great, great, great-granddaughter Valerie Bramwell and historian Robert M. Peck (Bramwell and Peck, 2008). Hawkins' existence was an eventful one, including both great professional success and tragedy, as well as a complex, sometimes unfortunate personal life. The latter is perhaps the most cryptic and unexpected part of the Hawkins story, as he somehow sustained nearly four decades of bigamous marriage to two wives, with whom he had ten children, seven of whom survived infancy. His unlawful marriages were found out in the mid-1870s, not long before a string of personal tragedies and a stroke brought Hawkins' life to a sad, inauspicious end. But despite his complicated family circumstances, Hawkins seems to have been a well-liked, generous figure known as hardworking, personable and charming, equally confident of his abilities as an artist and anatomist but also reverential and respectful to his peers, especially the academics and intellectuals he frequently worked with.

Hawkins' career was varied, especially early on, but his chief profession was creating art of animals – initially living, and eventually extinct. Known best today for the Crystal Palace models, Hawkins only began his sculpting career during the 1840s, well after he had established himself as an expert painter and illustrator. Even before the Geological Court he was regarded as a high-quality, experienced natural history artist and was in demand as an illustrator of zoological specimens. Across his career he produced art for many of the biggest names in nineteenth-century palaeontology and biology from the UK and USA, including William Buckland, Gideon Mantell, Richard Owen, Thomas Huxley, Charles Darwin and Joseph Leidy. He was sought after for his expertise in animal anatomy, which he learned primarily from zoological museum specimens and from drawing live animals, about which he would author several books. In later life Hawkins became an outspoken anti-evolutionist, delivering lectures and creating drawings outlining what he saw as major problems with Darwinian theory.

The Geological Court project was the acme of Hawkins' career and most of his post-1854 engagements were related to it in one way or another. He was the first artist requested to do the work by the Crystal Palace Company and, although his professional standing made his employment in this role unsurprising, the process by which he landed the job is unclear. Hawkins had already been successfully involved with the Great Exhibition, where he had served as District Superintendent and exhibited work, including a bronze-casted group of aurochs (his earliest known piece of palaeoart) and a model of horse anatomy. This, and other projects, would have made his work known to key figures in the Crystal Palace Company as well as royalty, such that Prince Albert (royal patron of the Crystal Palace) may have recommended him for creating the palaeontological sculptures. Alternatively, Hawkins was also highly regarded among leading academics – including Richard Owen – and his employment may have come at their suggestion. However it occurred, Hawkins was commissioned for the palaeontological component of the Geological Court in September 1852 and worked on their execution until April 1855: a pivotal period in his own career, the history of science communication, and the history of palaeoart.

Gideon Mantell (1790–1852), first-choice palaeontological consultant

Although Hawkins was a skilled natural history artist, the newly discovered, often poorly known fossil animals he was reconstructing for the Geological Court were a challenging study. A consultant to assist the design of the palaeontological sculptures was thus required to ensure Hawkins' work was as informed and up-to-date as it could be. This job would eventually go to Richard Owen (below), but first refusal for this role went to one of Owen's academic rivals, Dr Gideon Mantell (see Fig. 2.6c). Mantell is among the most famous and documented figures in early Earth sciences (e.g. Dean, 1999; Cadbury, 2000), chiefly for his pioneering work on *Iguanodon* and its significance to the early understanding of dinosaurs. But Mantell's academic career went far beyond this, including contributions to the study of many other extinct animals and British geology.

Although his professional life was one of mixed fortunes, by the early 1850s Mantell's status as a leading authority on prehistoric animals was not in doubt. Moreover, Mantell was an experienced public communicator through lecturing, authoring popular books and developing exhibitions – including a display for the 1851 Great Exhibition. He was thus an ideal candidate to work with Hawkins and was approached for this task in 1852. Mantell turned it down, however, on grounds that he disliked the contextless, specimen-free presentation of the Geological Court. Although Mantell at least tentatively endorsed the use of life restorations, he felt the

Fig. 2.7 Examples of Hawkins' excellent artistic skills and anatomical knowhow from the 1844–75 tome *The Zoology of the Voyage of HMS Erebus & Terror*, edited by John Richardson and John Edward Gray. Hawkins' understanding of natural history and experience working alongside leading scientists made him an ideal candidate for creating the palaeontological displays of the Geological Court.

28 Chapter 2 – Ancient worlds through a Victorian lens: planning the Geological Court

Fig. 2.8 Hawkins' lithograph *Skeleton of a man, with the skeleton of an elephant* (1860). Hawkins was a skilled artist who understood the anatomy of his living subjects, even when tackling large, exotic species like Asian elephants.

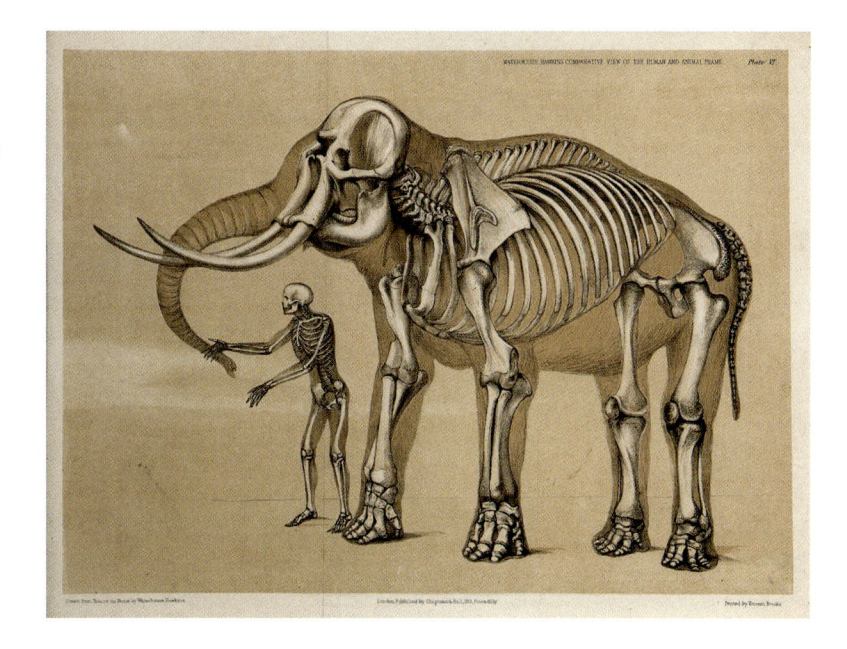

Fig. 2.9 Gideon Mantell was, like many early scholars, only tentatively on board with life restorations of extinct animals. Even his public-directed works, like his *Petrifactions and their Teachings* (1851), contain relatively few examples of restored extinct animals compared to illustrations of fossil specimens, and this conservative line art of *Megatherium americanum* is the only land vertebrate given this honour in this work.

Geological Court needed a more conventional fossil exhibition and explanatory panels alongside Hawkins' sculptures. The philosophy of the Crystal Palace project prohibited such features, however, and the two parties could not agree to work together.

A further factor in Mantell's refusal must have been his ill health. A carriage accident in 1841 had left Mantell with a terrible spinal injury that plagued the last decade of his life, leaving him physically disabled and in constant pain. He died of an overdose of opium in November 1852, just months after Hawkins began work. Mantell's palaeontological legacy is strongly evident in Hawkins' sculptures, however, especially in decisions taken in the reconstructions of the dinosaurs *Iguanodon* and *Hylaeosaurus*.

Richard Owen (1804–92), second-choice palaeontological consultant

When Mantell turned the Crystal Palace Company away, their second choice was Professor Richard Owen (*see* Fig. 2.6B), then conservator at the Royal College of Surgeons, London and shortly (in 1856) to become superintendent of the natural history department of the British Museum. Owen was a different character entirely to Mantell, as is well documented in accounts of the rivalry between these two pioneering figures (e.g. Dean, 1999; Cadbury, 2000; Rupke, 2009). Where Mantell was of modest social standing and performed his geological research around his medical practice, Owen was a member of the aristocracy, moving in the highest social circles and transferring from a medical career early in his professional life to exclusively pursue zoological and palaeontological research at prestigious institutions.

Owen became the preeminent comparative anatomist of his age and, in the latter half of his career, lobbied and oversaw the construction of a dedicated natural history museum for the British Museum's vast natural history collections: London's famous Natural History Museum was the result. His contributions to comparative anatomy, zoology and palaeontology are too many to list, but include seminal works on recent and fossil vertebrates of all kinds, and the recognition and naming of the reptile group Dinosauria. He is equally famous for debating Charles Darwin and Thomas Henry Huxley over evolutionary theory.

On paper, Owen was an ideal adviser for the Geological Court. He had worked with Hawkins previously (Gardiner, 1991; Bramwell and Peck, 2008), had progressive attitudes towards public education, was an undoubted authority on

Fig. 2.10 Richard Owen's contributions to natural history are many and earned him the reputation of being Britain's leading comparative anatomist during his long career. But he is most often remembered for his recognition of Dinosauria based on shared anatomical characters of *Iguanodon*, *Hylaeosaurus* and *Megalosaurus*, species that featured prominently at the Crystal Palace. (2015)

many of the species desired for the Geological Court, and his skills at anatomical correlation – the ability to predict whole body forms from isolated body parts (*see* below) – were legendary. He had attained great fame for his 1839 reconstruction of the moa (specifically, *Dinornis* – a giant, wingless, recently extinct bird of New Zealand) from a single leg bone (Dawson, 2016) and, based on this, he was surely the ideal consultant to help the Crystal Palace Company wrangle details of extinct animal life appearance from scrappy fossil remains. Owen's famed correlative feats gave the Crystal Palace Company a means to hype the authority of their project, portraying the famous professor as the intellectual director of the palaeontological restorations and casting Hawkins as their mere artist (Phillips, 1854). This culminated in Owen being charged with a second task, writing the dedicated visitor guidebook to the Geological Court: the 1854 *Geology and Inhabitants of the Ancient World*.

The perceived benefits of Owen collaborating with the Crystal Palace Company were not one-directional. In the mid-1800s Owen was engaged in several debates about the nature of fossil animals and vyed with Mantell for academic authority on extinct creatures, especially dinosaurs (Norman, 1991; Dean, 1999; Cadbury, 2000; Rupke, 2009; Torrens, 2012). As the first major public display on prehistoric life, the Geological Court presented him with a chance to shape popular opinion about the forms of extinct creatures and publicize his views at a scale and permanence unavailable to his rivals. Owen was similarly involved in discussions over the mechanisms by which life on Earth developed (Desmond, 1976), and the Crystal Palace models were an opportunity to strike a blow at early forms of evolutionary theory.

Owen did not follow creationist beliefs but his views on life's development involved some traditional Christian creation concepts, conflicting with pre-Darwinian evolutionary theory where life developed in a linear trajectory from less to more sophisticated organisms. These early ideas of biological evolution were growing into prominence throughout the 1800s and, for the ruling classes, conveyed a dangerous social message: if life itself could evolve, so could society. Owen's rapid professional ascendancy was based, in part, on his embracing of more traditional values as his friends and supporters saw political value in his rejection of radical evolutionary concepts (Cadbury, 2000).

To this end, Owen's research into dinosaurs has been cast as a directed attack on early forms of evolutionary theory: how could dinosaurs, which Owen demonstrated were reptiles with the 'advanced' qualities of mammals, be more 'complex' and sophisticated than living reptiles when they existed entirely before modern reptilian species (Desmond, 1976)? Having made this case in print, the Crystal Palace presented Owen with an opportunity to cast it in concrete. Such interpretations have coloured some scholarly perception of the Crystal Palace Dinosaurs, with Cadbury (2000) giving an especially charged take that casts 'Owen's rhinocerine dinosaur models' (p. 235) as 'monstrous gargoyles peeping out at the twenty-first century… a bizarre reminder of forgotten hopes and forgotten quarrels' (p. 236).

Behind the scenes, however, the relationship between the Crystal Palace Company and Owen was an uneasy one. From the start, Owen had issues with the contextless, specimenless approach desired by the Crystal Palace authorities and he echoed Mantell in desiring a more typical exhibition

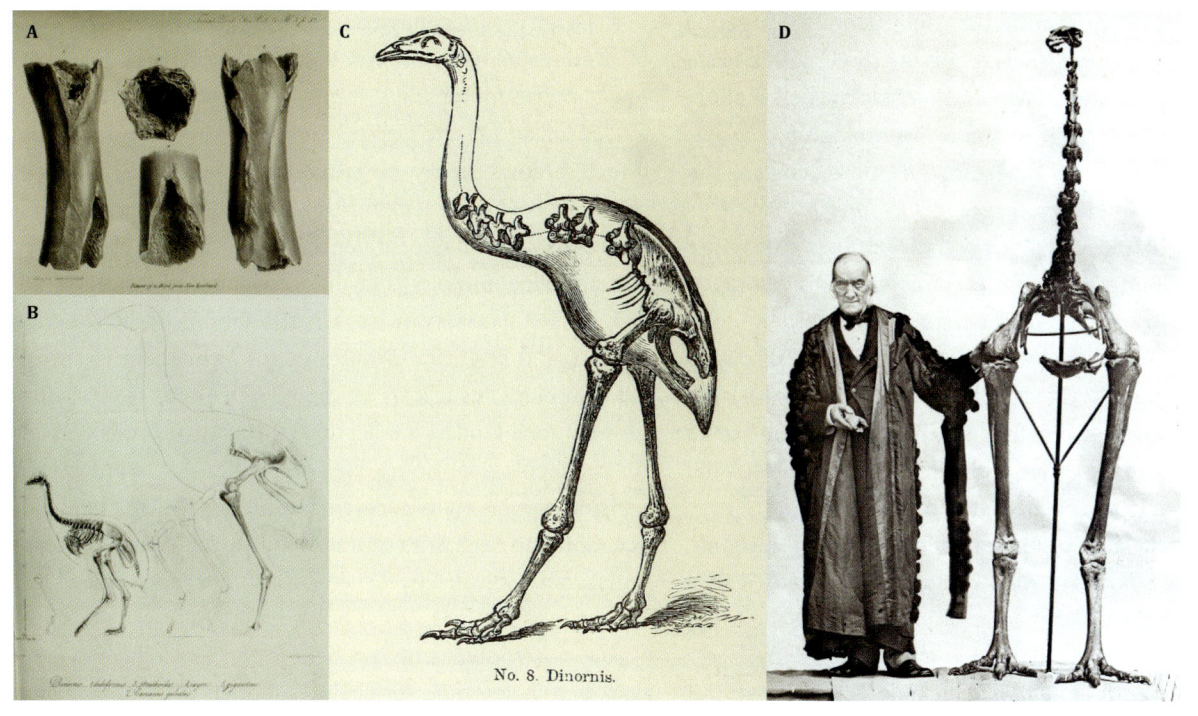

Fig. 2.11 Richard Owen and his most famous example of anatomical correlation: his prediction of the giant moa, *Dinornis*. **A)** the femoral fragment that Owen (ostensibly) used to predict the ostrich-like form of the moa; **B)** schematic drawings from Owen's (1842a) paper predicting moa form and size using slightly more fossil remains; **C)** skeletal reconstruction of *Dinornis* from the Geological Court guidebook (Owen, 1854); **D)** Owen stands triumphant with a complete *Dinornis* skeleton, with the original partial femur in his hand, in 1879.

alongside Hawkins' sculptures. He went so far as to organize a collection of fossil specimens for the Crystal Palace Company to purchase (Dawson, 2016). This deal never came to fruition, however, and it seems that Owen had little other involvement with the project in either an instructive or physical sense.

This can be contextualized against Owen's infamous professional conduct, which is as notable as his professional achievements. His academic brilliance came with an arrogance and disdain for his peers that has coloured his legacy and cast him, at best, as a complicated historic figure or, at worst, as a malicious, vindictive 'villain' of nineteenth-century science. Some scholars (Padian, 1997; Rupke, 2009) have rightly argued that Owen has been demonized and misrepresented in his disputes with figures like Mantell and Charles Darwin, but it is also an inescapable fact that he was fierce and sometimes cruel to his peers, that he exploited discoveries made by others, and that he frequently rewrote history in his favour (Cadbury, 2000). Such professional conduct was already damaging his reputation in the mid-1800s and, while there is no evidence he was cruel or callous to anyone involved in the Crystal Palace project, his lack of interest and seeming unwillingness to assist in his consultancy role is now well documented (Secord, 2004; Dawson,

2016), despite his later efforts to claim that he was first choice as technical advisor to the project as well as the man who suggested hiring Hawkins (Dawson, 2016).

That Owen was an absent consultant is evidenced in several ways. Most glaringly, letters to Owen from Hawkins and the Crystal Palace Company repeatedly requested his attendance during construction, but visits were few and probably limited to special events, such as the famous 1853 New Year's Eve *Iguanodon* banquet held in his honour (Secord, 2004; Dawson, 2016). Newspaper reports reveal that Owen only saw the models after they were installed in the Geological Court, leading Owen to defend parts of reconstructions he had not authorized (Chapter 7; Secord, 2004). Even Hawkins, who clearly held his consultant in high regard and had good relationship with Owen in general (Secord, 2004; Bramwell and Peck, 2008), wrote a frank and honest assessment of Owen having 'afforded no assistance' during the development of the models (Dawson, 2016). Owen's absence probably explains why several of Hawkins' restorative decisions follow the works of other scholars and, in turn, why Owen was sometimes dismissive of Hawkins' choices in his Geological Court guidebook.

Owen also took shortcuts with his other major commitment to the Geological Court, the official guidebook.

Such books were necessitated by the intentional absence of informative signs or panels alongside the Crystal Palace exhibits and, while general park guides provided a short description of the geological and palaeontological displays, Owen's contribution was expected to be more detailed. The result, *Geology and Inhabitants of the Ancient World*, is probably Owen's most significant contribution to the entire Crystal Palace enterprise, but is notably slim, incomplete, and replete with errors. It provides a concise overview of the production of the palaeontological restorations and discusses the sculptures and recreated geology of the Secondary Island, but not a single reference is made to the Tertiary Island or its inhabitants, nor to the major geological displays representing 'Primary' strata in the northwest of the Court.

A *Dinornis* skeleton is illustrated in the back of the book, but readers are not informed of its significance. It could be a reference to Owen's association with the moa and anatomical correlation or a nod to the intention to place a *Dinornis* model in the Geological Court (Chapter 12), but it is ultimately a contextless inclusion that seems to have been an afterthought or remnant of an unfinished draft. Alongside referencing errors and digressions about matters unrelated to the Crystal Palace, there are many reporting errors related to the Geological Court displays.

Just some examples include giving the length of the '*Plesiosaurus*' *macrocephalus* sculpture as 18ft (5.5m) long and 7ft (2.1m) wide (the model is 20–30% smaller than this); the suggestion that only the head of the *Mosasaurus* was reconstructed (much of the body and one flipper was also made); and that one *Dicynodon* species was only 'partially restored' (the whole animal was built). He also undermines the authority of Hawkins' restorations on several occasions, noting that, for example, *Iguanodon* having a nose horn was 'more than doubtful' (p. 17) and that the form of the '*Labyrinthodon*' was 'more or less conjectural'. Dawson (2016) reports that Owen reportedly didn't even proofread his own text, delegating the responsibility to Hawkins. Although not totally lacking merit – it contains some fine and historically significant illustrations, such as the first skeletal reconstruction of a dinosaur (Chapter 7), and Owen's text gives a reasonable insight into certain aspects of palaeontological science of the 1850s – *Geology and Inhabitants of the Ancient World* is mostly notable for its shortcomings.

In light of these facts, Owen's role with the Crystal Palace Company is being reconsidered. Ideas that the Crystal Palace Dinosaurs were exploited as an Owenian political project to stake victory in his feuds with Mantell (e.g. Norman, 1991; Dean, 1999; Cadbury, 2000; Rupke, 2009; Torrens, 2012) or undermine early evolutionary theory (Desmond, 1976) must now be doubted, as Owen's disinterest is simply too well evidenced to see the Crystal Palace Dinosaurs as carefully directed blows against his rivals. Any Owenian qualities of the palaeontological sculptures are probably more a reflection of Owen's inescapable influence in mid-nineteenth-century British palaeontology than his direct consultancy. While this in itself has relevance to interpretations of the complex politics and scientific debates of mid-nineteenth-century palaeontology, it suggests we should move away from casting Owen as any kind of intellectual authority or project leader for the Geological Court. He was, at best, an uninterested consultant who mainly served as a prop for the Crystal Palace Company to claim authenticity in their palaeoartistic endeavour. Conversely, it seems that Waterhouse Hawkins, long cast as the mere artist bringing life to Owen's designs, was a more potent intellect behind the Geological Court's palaeontological models than has generally been appreciated.

Conceptualizing the past

With a full complement of intellectual, engineering and artistic talent recruited, attention turned in late 1852 to the mammoth task of designing and building acres of recreated geology, landscaped lakes and islands, and over thirty sculptures of extinct animals. Documentation of this planning phase is patchy, being relatively good (though far from complete) for Hawkins' prehistoric animals but virtually non-existent for Ansted and Campbell's geological features.

We know that Ansted collaborated with Paxton to ensure the Court layout was congruent with wider plans for the Crystal Palace grounds (Phillips, 1856) including, it must be assumed, the complex scheme of artificial tides associated with the giant park fountains (Chapter 3). We also know that Ansted set out to create a representation of major British geological strata, including beds of economic importance as well as those related to spectacular fossil discoveries, and that he wanted to arrange them accurately to convey their position in geological time (Doyle and Robinson, 1993; Doyle, 2008). Beyond this, our understanding of Ansted's plan becomes less certain and we are best informed of his vision by analysing the Geological Court itself, which we will do in the next chapter.

More can be said of the plans that went into Hawkins' and Owen's palaeontological component. Archived letters, press

Fig. 2.12 Illustrations of the Geological Court featured in Crystal Palace guidebooks (examples here from Phillips 1856 and Owen 1854). **A)** view of the Secondary Island and dinosaurs; **B)** the trio of *'Labyrinthodon'*; **C)** pterosaurs, perhaps *'Pterodactylus' cuvieri*; **D)** the Secondary Island, showing most of the Mesozoic species in approximately correct positions; **E)** the curiously sparse map from Owen's 1854 guidebook, which not only omits most of the sculptures and Geological Illustrations, but presents a confused overview of the geological arrangement.

articles and lecture transcripts directly relate to the design phase of the prehistoric animal sculptures, many of which tie into well-understood wider developments of palaeontological theory in the nineteenth century. Combined with data taken from the finished models, these records allow us to produce a reasonable insight into the scientific philosophy and palaeoart processes that underlie Hawkins' sculptures.

A core principle of their execution was that they would not, as was common in other artwork of extinct animals in the early nineteenth century, simply replicate restorations that had come before. They would be wholly original designs incorporating the latest science to recreate familiar species anew, and restore others to life appearance for the first time, capitalizing on Hawkins' knowledge of rendering animal anatomy and Owen's knowledge of extinct organisms. The sculpture team accordingly employed an approach that would be familiar to modern artists of prehistoric animals. As summarized by Owen (1854):

> Those extinct animals were first selected of which the entire, or nearly entire, skeleton has been exhumed in a fossil state. To accurate drawings of these skeletons an outline of the form of the entire animal was added, according to the proportions and relations of the skin and adjacent soft-parts to the superficial parts of the skeleton, as yielded by those parts in the nearest allied living animals.
>
> OWEN, 1854, PP. 5–6

This passage exaggerates the fossil information available to Hawkins and Owen as many species restored at Crystal Palace were far from being represented by entire skeletons, but the outlined methodology of reconstituting fossil bones into a full, realistically-posed skeleton, following guidance from living species to add muscle and bulk, and then clothing the reconstructed species in an appropriate skin, is exactly the same process used by palaeoartists today (Witton, 2018). We know that Hawkins visited specimens of his subject species in person to measure their bones and understand their anatomy as well as he could before commencing work (Hawkins, 1854).

The original vision of the Geological Court was to focus on mammalian subjects from Tertiary deposits, with the mastodon being the first planned sculpture (Hawkins, 1854).

Had Hawkins followed this plan, his research and planning phase would have been far easier: many of the large, impressive Tertiary species he would have created (plans included mammoths, mastodons, big cats and others) were represented by excellent fossil material that left little doubt about their size, proportions and general osteology. Moreover, in being no strangers to modern mammal anatomy, both he and Owen would have found recreating such species a relative breeze: no more evidence of this is needed than the excellent, and still essentially accurate, Geological Court mammals (Chapter 4).

But mammalian subjects were sidelined after Hawkins read research papers on the then newly discovered saurians emerging from European deposits, including dinosaurs, marine reptiles, and pterosaurs (McCarthy and Gilbert, 1994): he was suitably enamoured with these spectacular reptilian subjects that their design and construction became prioritized. But they also presented a complication to Hawkins' task, as many were known from scant fossil material. Without complete osteological blueprints to follow, Hawkins and Owen would have to imagine, predict and speculate swathes of unknown anatomy to bring these animals to life. How could this be done while still maintaining a vestige of scientific authority?

The answer to this question lay in the philosophy of anatomical correlation. This idea has origins with the grandfather of comparative anatomy, Baron Georges Cuvier (1769–1832), who predicted that the forms of entire animals could be reconstructed with some degree of precision from very few anatomical remains. Cuvier's concept, explored in depth by Gowan Dawson in his 2016 book *Show Me the Bone*, had gained considerable traction among anatomists as Cuvier used his knowledge of animal anatomy to predict the forms of poorly known fossil Palaeogene mammals of the Paris Basin. Whereas fossil species like *Megaloceros*, mammoths and *Megatherium* had been discovered from relatively complete skeletons, the fossil mammals of Paris were often unearthed as isolated bones and partial skeletons. Cuvier's knowledge of comparative anatomy allowed him to identify characteristics of these fossils that informed both their taxonomic allocation as well as the form and shape of the whole animal. He spoke of being capable of making such deductions from solitary bones:

Fig. 2.13 Among the many challenges facing Hawkins with his Crystal Palace commission was creating a vision of prehistoric life that would be convincing to a public largely naive to the existence and nature of extinct organisms. This may not be so difficult with fossil mammals and crocodylomorphs, but creatures like *Hylaeosaurus* had not been reconstructed in detail before, let alone at life size, and never for public view. (2017)

Fig. 2.14 C. L. Laurillard's famous restorations of the mammals of the Paris Basin drawn under the guidance of Georges Cuvier, featuring *Anoplotherium commune*, 'Anoplotherium gracile', 'Palaeotherium minus' and *Palaeotherium magnum* (from Cuvier, 1822). These restorations were created using the Cuvierian principle of anatomical correlation from various grades of fossil remains, and also informed by detailed muscle studies (unpublished in Cuvier's lifetime). The palaeontological restorations at Crystal Palace would be approached using the same philosophy.

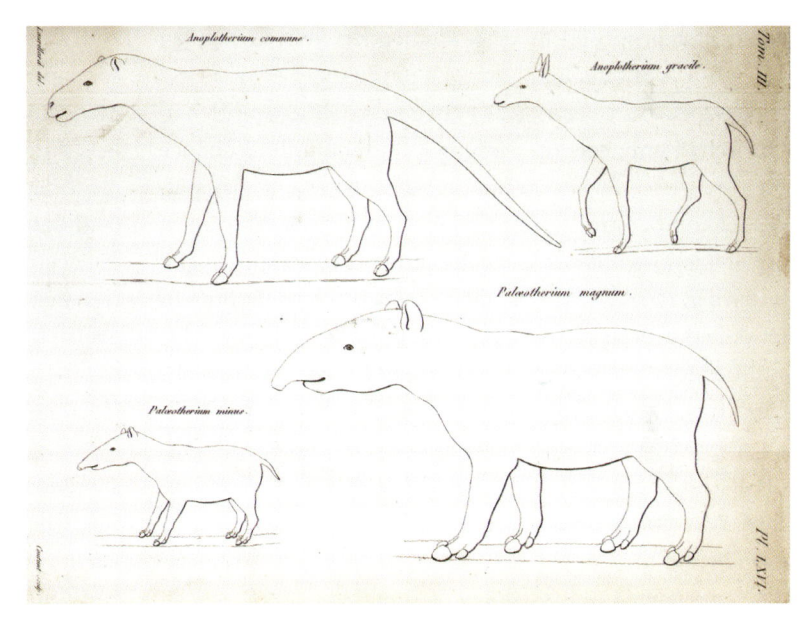

Today, comparative anatomy has reached such a point of perfection that, after inspecting a single bone, one can often determine the class, and sometimes even the genus of the animal to which it belonged… The bones that compose each part of an animal's body are always in necessary relation to all the other parts, in such a way that – up to a point – one can infer the whole from any one of them.

CUVIER, 1798

(QUOTED IN DAWSON, 2016, PP. 25–26)

The idea that palaeontologists can infer the entire form of an animal from solitary bones and bone fragments is naive against a modern understanding of biodiversity through Deep Time, but Cuvier's fundamental observation that the features of bones can be diagnostic to specific animal types and broader aspects of form remains true. It was this principle that allowed early anatomists to find order in imperfect fossil assemblages and not, as happened with fossil reconstructions before Cuvier, create chimeras of different animal types (e.g. the 1663 *unicornum verum* – a blend of Pleistocene mammal bones forged into a fantastical narwhal-like creature). Although some scholars – including Mantell – developed misgivings about anatomical correlation relatively early in the 1800s, it played an important role in interpreting and classifying fossil life in the early phases of palaeontological research.

Post-Cuvier, Owen became one of anatomical correlation's biggest proponents. His ability to recreate extinct creatures from scrappy bones was celebrated in Victorian culture, even attracting the attention of royalty (Rupke, 2009; Dawson, 2016). Owen's 1839 prediction of the form of *Dinornis*, the giant moa, from a fragment of limb bone was the most lauded example of his predictive prowess. This tale is generally recorded as Owen being presented with a six-inch long mid-shaft of a limb bone from New Zealand that had confounded other scholars, but was generally identified as pertaining to a large bovid (*see* Fig. 2.11A). Owen's anatomical skills identified structures which foresaw something entirely different, however. From the fragment alone, he managed to reconstruct the outline of a whole femur, which he saw as indicating 'a struthious bird nearly, if not quite, equal in size to the Ostrich' (*see* Fig. 2.11B; Owen, 1840a, p. 170–71).

Considered an audacious claim by Owen's colleagues when first proposed in 1839, vindication arrived in 1843 when boxes of moa fossils arrived in the UK. New Zealand had, indeed, once been home to large, ostrich-like birds, and Owen had only needed a fragment of hindlimb to deduce their existence. It was a remarkable achievement that suggested comparative anatomy was entering a new realm of understanding where fossil material, however scrappy, could be used to reliably and exactly reconstruct whole animals by skilled experts. Owen's influential friends promoted his abilities to a near-legendary status and the Crystal Palace Company, keen to emphasize the scientific integrity of their Geological Court project, made good use of Owen's *Dinornis* prediction in their promotional activities. Any criticism of their reconstruction authenticity – which came their way even before the Court was finished (Chapter 12) – could be deflected using Owen's prior achievements.

In truth, Owen's prediction about the moa was exaggerated and over-hyped, a fact consciously ignored and hidden by those with a vested interest in Owen's reputation (Rupke, 2009; Dawson, 2016). Indeed, Owen was not even the first to identify the moa bone fragment as avian. As early as 1848 it emerged that the naval surgeon Dr John Rule, who had brought the limb fragment to Britain, had identified avian features in the specimen and communicated them to prospective buyers – including Owen. Rule never achieved the recognition he deserved for this in his lifetime because Owen and his supporters selectively omitted his identification, and eventually even his name, from retellings of the moa story (Dawson, 2016). Moreover, the accuracy of Owen's 'correlation' was over-sold. Although undoubtedly correct in anticipating the femoral shape and basic ostrich-like form of the moa, Owen had not been able to predict their other defining characteristics, such as their wingless bodies and legs adapted for slow locomotion. He also underestimated the size of *Dinornis* by some margin: giant moa could be up to 3.6m tall, almost twice the height of the ostrich-sized bird Owen envisaged. It is thus fair to regard Owen as having deduced some basic essence of the moa, but he had not – as some of his proponents claimed – conjured an exacting vision of this extinct bird from a scrap of bone.

The reality behind Owen's moa work had little impact on the popularity of anatomical correlation in the early 1800s, however, and it would serve as a philosophical cornerstone of the reconstruction process employed for the Crystal Palace Dinosaurs. It served as a technique to predict the form and size of any species known from incomplete remains, firstly by Owen, who made appropriately conservative deductions about creatures such as dinosaurs, *Dicynodon*, *Mosasaurus* and '*Labyrinthodon*' from scrappy fossil data in his technical papers and notes, and then secondly by Hawkins, who made further inferences about the same species to fill gaps in Owen's deductions. Hawkins' inferences were necessarily more speculative than Owen's (he had, after all, to present completed reconstructions rather than general concepts) but were not mere artistic whimsy: they were informed, reasoned speculations of form and function that imagined plausible anatomies in places where palaeontologists had yet to provide insights.

Examples include his inference that the heavy heads of *Iguanodon* and *Megalosaurus* must have required bulging shoulder regions to anchor skull-supporting musculature (*see* Chapter 7) and his suggestion that large pterosaurs would, like modern birds, have proportionally larger torsos, and thus larger flight muscle volumes, than smaller ones (Chapter 6). Hawkins also attempted Cuvierian or Owenian feats of whole body prediction using nothing but partial skulls and a single vertebra to predict a turtle-like body for *Dicynodon* (Chapter 11).

We may view Hawkins' predictions as naive today and as further evidence that, as with Owen and his moa, prehistoric animals cannot be exactly reconstructed from scraps of bone. But when considered through an appropriate historic context, they demonstrate a sound, logical approach to extinct animal restoration that was well ahead of other artists of the time. Hawkins' application of anatomical correlation and form-function relationships established a precedent for using informed inference and deduction in palaeoart that would typify palaeoart practices for years to come, and rise to particular popularity in the early twenty-first century 'All

Fig. 12.15 Hawkins evidently had a clear concept of the anatomy of his creations even when the necessary fossils had not been found. This is evident from the body contours of reconstructions like his *Iguanodon*, which clearly correspond to a fully mapped out muscle system. This is a key reason why his reconstructions were both realistic and ahead of many of his contemporaries; even though his finished products were scientifically flawed, they look like animals that could actually exist.

Yesterdays' movement, where palaeoartists employ reasoned speculation to explore 'known unknowns' of prehistoric animal anatomy (Conway *et al.*, 2012). As we will explore in later chapters, many of Hawkins' predictions anticipated genuine aspects of prehistoric animal anatomy even where the overall forms of his reconstructions were incorrect.

Methodological reconstruction processes and anatomical correlation were not the only techniques used in developing the Crystal Palace prehistoric sculptures. Hawkins also took guidance and inspiration from living animals, sometimes even transplanting entire components – heads, bodies, skin textures and so on – to his creations. The application of modern anatomy to extinct animals is also a widely used technique among modern palaeoartists and, as is seen at Crystal Palace, is a practice that cuts both ways.

A certain amount of anatomical transfer is inevitable in palaeoart and is often appropriate. Extinct animals are often, for example, closely related to living ones and likely shared some common anatomy. Even when this is not the case, sensibly applied and well-rationalized transference of modern anatomy can make strange and unfamiliar extinct species seem more grounded and relatable. Anatomical transference is also unavoidable when we have no fossil evidence to guide us any other way, and Hawkins would have been especially reliant on modern animal anatomy in the 1850s when so much about his extinct subjects had yet to be discovered. He would have had little choice other than transferring the skins of living reptiles and amphibians to his dinosaurs and '*Labyrinthodon*', for instance, and his reconstruction of *Palaeotherium medium* as a variant on a modern tapir would

Chapter 2 – Ancient worlds through a Victorian lens: planning the Geological Court have seemed scientifically sound, it being thought at the time that these creatures strongly resembled tapirs in life (Rudwick, 1997).

But in some instances Hawkins seems to have overwritten fossil data with the anatomy of living species, such as his application of modern crocodylian armour to *Teleosaurus* (*see* Chapter 8) or monitor lizard facial features to the *Mosasaurus* (Chapter 5). These errors are especially curious in light of Owen's purported involvement with the Crystal Palace project: in several instances Owen had first-hand experience with the fossil data that Hawkins overrode with anatomies from living animals, but the mistake was not prevented. The precise circumstances behind these errors are unknown, but they are consistent with Owen being a largely absent advisor during the planning and execution of the models.

In addition to anatomy, Hawkins also had to decide on poses, demeanour and behaviours for his recreated menagerie. Little is known on the thought processes behind these decisions, although the anonymous 1854 *Hogg's Instructor* article *The Geology of the Crystal Palace* demonstrates that some of Hawkins' early plans differed vastly from the completed models. They once included wrestling plesiosaurs and pterosaurs engaged in acts of flying, running and diving, compositions perhaps consistent with those nineteenth-century palaeoartworks that gave prehistory a violent, dark tone (*see* Fig. 1.9). For uncertain reasons, Hawkins steered the project in a more naturalistic direction by the time the sculptures were put into production. The models are posed in relatively stately manners that imply movement – walking, emerging from water, reclining, rearing, stretching or displaying – without strongly imparting any particular mood or judgement of their place in the order of history. Intentional or not, Hawkins' choice of behaviours and poses imbues his creations with the same unreadable, mysterious intelligence that we associate with many living animals: a quality that allows visitors to draw their own conclusions about their intent and state of mind.

This was not lost on journalists visiting Hawkins' workshop. In a century where prehistoric animals were often described and depicted as malign and violent, a *Hogg's Instructor* journalist regarded the half-built *Iguanodon* as one might view a cow or horse:

> Mr Hawkins has given the monster a somewhat mild and gentle aspect, however; and, as it stands with its great paunchy mouth partially opened, disclosing its curious leaf-shaped teeth, it seems a very model of innocence and contentment – a huge antediluvian illustration of the amiability produced by a strictly vegetarian face.
>
> ANONYMOUS, 1854, P. 285

The ability to imbue concrete and metal prehistoric animals with realistic, grounded behaviours is aided in several instances by contextualizing variably posed animals of the same species in small areas (*Megaloceros*, *Anoplotherium*, *Pterodactylus*, *Iguanodon*, '*Labyrinthodon pachygnathus*'), or arrangements of mixed species from the same geological horizons. This diorama-like approach mimics real-life encounters with wild animals in drawing our attention to not only the single sculptures but also their relationship to each other and their environment. It is among such a display that we see one of Hawkins' most progressive depictions of extinct animal behaviour: an *Anoplotherium commune* shaking itself dry as it emerges from water. Early palaeoart largely focused on depicting fossil animals in either purely illustrative fashions (*see* Figs 2.9, 2.14) or engaging in aggressive behaviour mostly related to predator-prey relationships (*see* Figs 1.8 and 1.9). The decision to show *Anoplotherium* shaking itself dry – a transient, comfort-based act specifically tied to the hypothesis of *Anoplotherium* as a swimmer (*see* Chapter 4) – brought hitherto unreached levels of nuance and insight into depictions of fossil animals. Indeed, such recreations remained rare for the next century as illustrative, trophic- or aggression-based depictions dominated palaeoart well into the twentieth century.

CHAPTER 3

Bricks, iron and tiles: rebuilding the past

Constructing the Geological Court and its content was such an enormous undertaking that it vastly overshot the development and opening of the other Crystal Palace attractions. Books and newspaper articles printed after the opening of the park in June 1854 often mention that finalization of both geological and palaeontological features were still ongoing, and it was only in April 1855 – almost a year after the Crystal Palace opened to the public – that *The Observer* reported Hawkins installing his final sculpture, one of the Cretaceous pterosaurs.

That work on the Geological Court overran is not surprising: the amount of planning, landscaping, plumbing, building and sculpting it required was tremendous. Photographs and illustrations from the construction period capture the remarkable effort required to landscape the grounds, craft the models and install the geological displays. These feats were all the more remarkable when contextualized against the similarly vast build of the Crystal Palace, its exhibitions and gardens; this was just one of dozens of similarly spectacular displays being built in the early 1850s on the southern slope of Sydenham Hill. Of course, this huge amount of work came at great expense: the Geological Court alone cost £13,792 (McCarthy and Gilbert, 1994; equivalent to £1.5 million when adjusted for inflation in 2021). It was this cost that ultimately stopped further development of the site and left the Tertiary Island geology and palaeontological displays only partially realized (Chapter 12). In this sense, we cannot really speak of work on the Court being 'finished' in 1855 so much as 'stopped'.

Our understanding of how the Geological Court was built is incomplete. Much can be learned by analysis of the surviving sculptures, Geological Illustrations and landscaping, and we can find many details of their construction committed to record in contemporary newspaper articles, lectures and magazine pieces. But while documentation of the palaeontological restorations is relatively good, few details of the creation of the geological displays are currently known because they never generated the same level of interest as their extinct animal counterparts. Any question of how the Geological Court landscape was realized can, therefore, only be partially answered.

Constructing the Geological Court and the Geological Illustrations

Today, little mention is made of the restored geology of the Crystal Palace Park even though it remains – courtesy of extensive restoration – just as prominent in the park as the mammal and reptile sculptures. At first glance, it may seem that the rocks and cliffs associated with the palaeontological

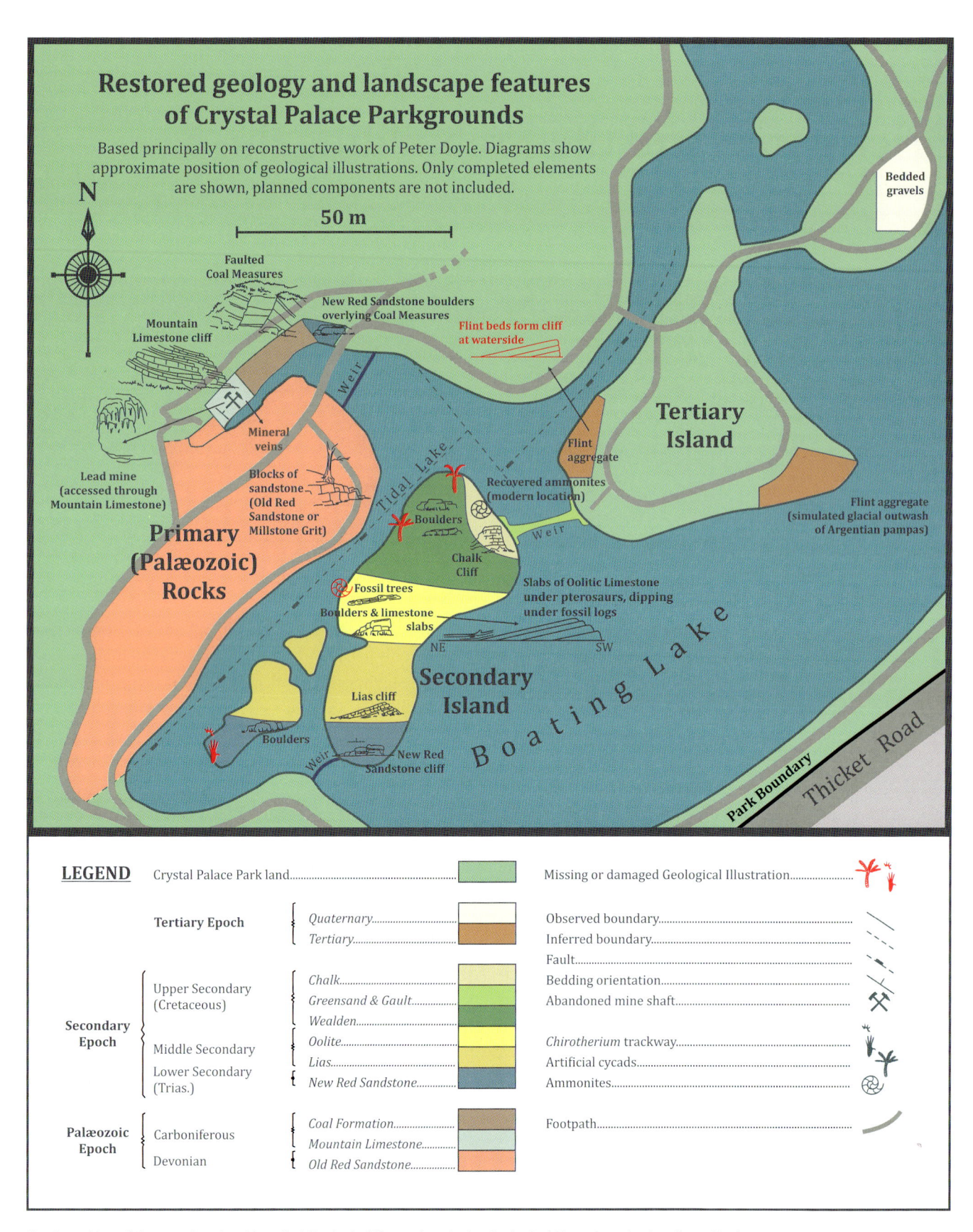

Fig. 3.1 Map of the completed and installed Geological Illustrations in the Geological Court, based primarily on Doyle and Robinson (1993) and Doyle's conservation notes from 1995, as well as new studies.

40   Chapter 3 – Bricks, iron and tiles: rebuilding the past

sculptures are mere stage dressing – an attempt to make a leafy London park look more 'prehistoric' – but they were actually constructed as an attraction and educational display in their own right. Their layout was designed to immerse visitors in the concepts of Deep Time, the law of geological superposition (see below), the basic arrangement of British geology and the industrial applications of earth sciences. The geological displays integrated with Hawkins' sculptures to present a virtual 'walk through time' where artificial rock outcrops and simulated geological features led guests from the oldest strata to the youngest, showing characteristic rock types, examples of their fossil content, and the industrial applications of geology along the way. Among the features of this complex simulated geological arrangement were a faulted coal face, several large cliffs, and even a lead mine in an artificial cave.

The aims of the geological restorations were not entirely aligned with the palaeontological sculptures nor, for that matter, the restorative principles of the rest of the park. Hawkins and Owen employed their imaginations as much as their scientific knowledge to create flesh recreations of animals now lost to time, just as other Crystal Palace teams sought to rebuild ancient Egypt and Rome as they appeared in their heyday, not as crumbling relics (see Fig. 2.4). Following these examples, Ansted and Campbell might have created diorama-like backgrounds for the palaeontological sculptures with appropriate ancient plants and landscaping, but they instead replicated modern geological outcrops, even basing their displays on real UK field localities.

This relatively conservative approach created a complex interpretative message for the Geological Court's visitors, mixing 'uninterpreted' science (replicated details of British geology) with highly interpreted scientific speculation (animals reconstructed from fragmentary fossils). While this may have been confusing to some, grounding the Geological Illustrations in conventional science may have been an attempt to ward off concerns about the scientific integrity of the palaeontological interpretations. Such criticisms surrounded the development of the Geological Court from its conception (Chapters 2 and 12) and the inclusion of straightforward geological displays alongside more speculative animal models may reflect efforts to infuse hard science into an exercise facing charges of being too fantastical.

Whatever the theory behind them, the choice to base the Geological Illustrations on real outcrops is surely one reason they have largely been ignored by all but the keenest historians; these are simply not seen as exciting to non-specialists as their palaeontological counterparts. Accordingly, little documentation seems to have survived about their history or subsequent modifications by later park custodians. Plans and design sketches are unknown (Doyle, 2008) and they receive only partial attention in Owen's (1854) Geological Court guide. Ansted and Campbell are not even named. Documentarians working to capture the building of the grander Crystal Palace project also gave no specific attention to the geological features and, with a few exceptions, little interest was paid to the recreated geology in the press.

Accordingly, most of our understanding of the construction and even the layout of the Geological Court has been achieved retrospectively, based on brief passages in general Crystal Palace guidebooks (e.g. Phillips, 1854; McDermott, 1854) and modern investigations of the court itself, especially by geologist Peter Doyle, whose 'geological mapping' of the Crystal Palace site has proven especially valuable (Doyle, 2008; Doyle and Robinson, 1993, 1995). Doyle has also been a champion for restoring the Geological Illustrations to the same level as their palaeontological counterparts, and their rescue from complete degradation in the late twentieth century owes much to his efforts. Unfortunately, the geological features have proven less sacrosanct than the palaeontological sculptures and several major changes have been made to their layout over time, including the complete disassembly of some sections. This adds an additional level of complexity to interpreting how the court was constructed and arranged, as some of the original layout can only be inferred or guessed from archaeological analysis.

Ansted's plan to have visitors walking through time is more accurately described as visitors walking along a series of differently-aged rocks, with those at one side of the court geologically younger than those at the other. To achieve this, Ansted had to carefully plan the arrangement of his simulated strata to ensure they accurately reflected real stratigraphic relationships – in other words, that the rock layers were arranged in geochronological order. As an educator and textbook author as well as a geologist, Ansted was familiar with distilling the complexities of geological phenomena into accessible models and, indeed, some components of the Geological Court are very similar to diagrams in his 1856 book, *Geological Science: Including the Practice of Geology and the Elements of Physical Geography*.

The Court was divided into three groups of strata representing the Primary (Precambrian – Palaeozoic in modern terminology), Secondary (Mesozoic) and Tertiary (Cenozoic), with the 'outcrops' in the northeast being younger than those of the southwest (see Fig. 3.1). Although some terminology differs, the Geological Column of 1856 is

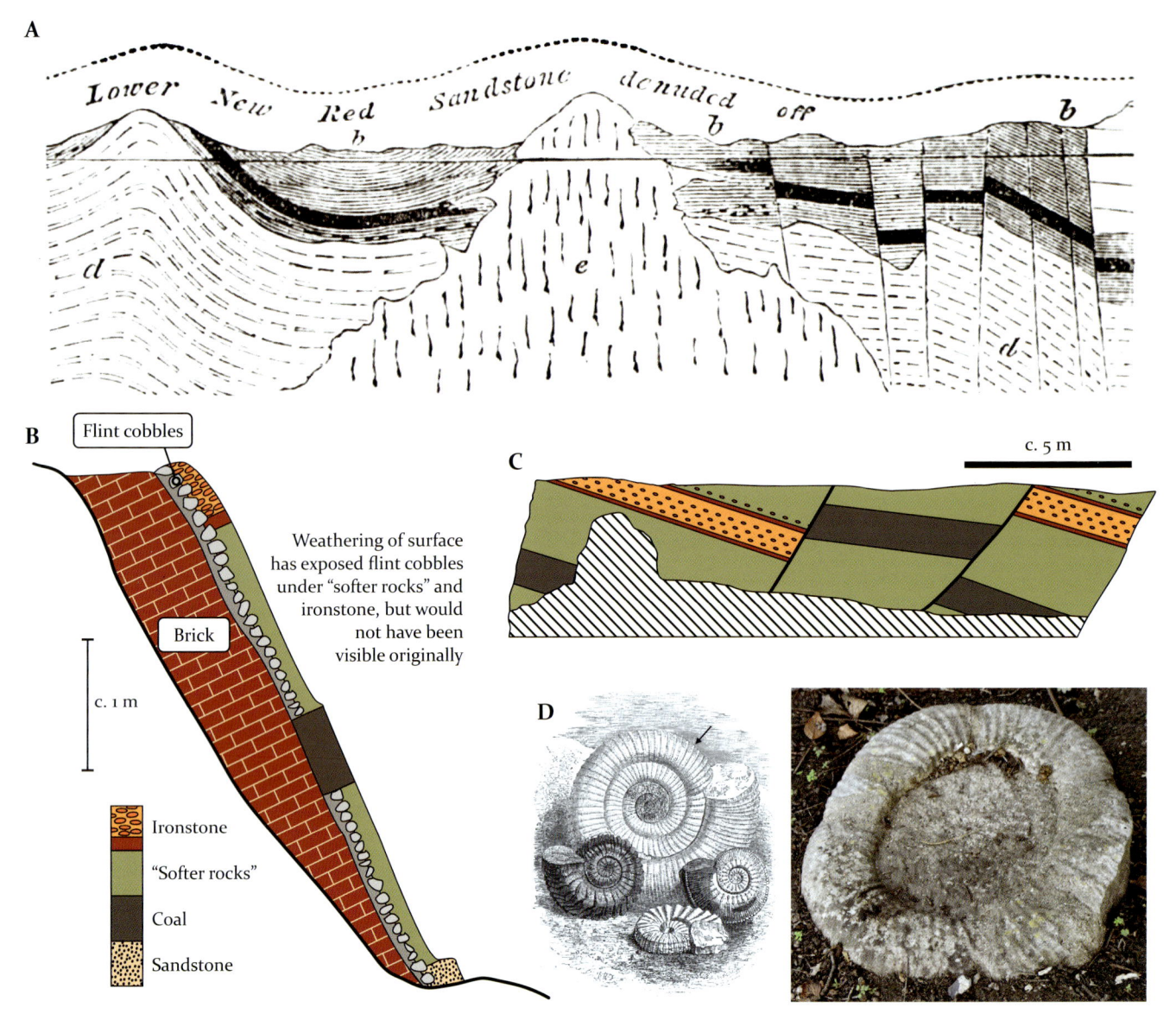

Fig. 3.2 Concepts and construction of the Geological Illustrations. **A)** diagram of Primary strata from Ansted's *Geological Science: Including the Practice of Geology and the Elements of Physical Geography* (1856), which strongly resembles the Primary outcrops of Crystal Palace. It shows faulted Coal Measures (*b*), underlying and older stratified rock (*d*, with obvious similarity to the Mountain Limestone), and an intrusive rock (*e*), meant to be basalt or porphyry, resulting in jointing and vein formation that could house lead. An unconformable cap of New Red Sandstone overlies the sequence, as at Crystal Palace; **B)** construction of the coal seam outcrop (modified from Doyle and Robinson, 1993); **C)** 'field sketch' of the coal seams (legend as in B) based on Doyle and Robinson (1993); **D)** *Titanites* ammonite fossils of the Geological Court, showing actual *Titanites* appearance (arrowed) vs the badly restored ammonite fossils present in the park today. Ammonite line drawing from *El Mundo Ilustrado* (1879).

surprisingly similar to that of modern times, with the most obvious difference being the absence of estimated ages of different eras, periods and epochs. Although some notion of the magnitude of Deep Time had been gleaned by the mid-1800s, accurate determination of the Earth's age was still a century and several major scientific discoveries away. Ansted wisely avoided any controversy of endorsing early, erroneous ideas on these matters by sidestepping the issue entirely.

Instead, Ansted focused on the core principles of mid-nineteenth-century geology: the nature of different rocks, the fossil organisms they contained, and their relative ages to one another as determined by laws of superposition: the axiom that, in undeformed sedimentary rock, older rocks and sediments were deposited below younger rocks. These would be demonstrated using strata recreated from the British rock record so that the Geological Illustrations conveyed

Geological Column 1850s *sensu* Ansted 1856

Age (My)	Epoch	Series	Ansted's list of included strata (bold text indicates representation in Geological Court)
Absolute dates unknown	I. Tertiary Epoch	Superficial Deposits	Raised Beaches - peat bogs - submerged forests - mud deposits in caverns - shell marls and modern deltas.
		Upper Tertiary	Gravel beds - **Till** - mammaliferous crag of Norfolk - Red Crag - upper limestones of Sicily - **subapennine beds** of Northern India (Kunkar, &c), **South America**, Australia, and other countries.
		Middle Tertiary	Coralline Crag - Upper Molasse of Switzerland - crag cliffs of the Loire and Garonne - Tertiaries of Vienna - numerous beds in India and America.
		Lower Tertiary	London and Hampshire clays and sands - Isle of Wight beds - **beds of the Paris basin** and Brussels - Lower Molasse of Switzerland - lower beds of Sewalik Hills, India - nummulite of other limestones of Eastern Mediterranean.
	II. Secondary Epoch	Upper Secondary or Cretaceous	1. **Chalk of England, France, Belgium, and Denmark**, *Seaglia* of Italy 2. Lower chalk and chalk marl, Quadersandstein of Germany 3. Upper Greensand or Firestone 4. Gault 5. **Lower Greensand** or Neacomian
		Middle Secondary	*a. Wealden Series.* 1. Weald Clay 2. Hastings Sand 3. Purbeck Beds. *b. Oolitic,* or *Jurassic Series.* 1. Portland Beds, and lithographic beds of Bavaria 2. Kimmeridge Clay 3. Coral Rag, and Nerinaean Limestone 4. Oxford Clay - 5. Cornbrash, Forest marble, Bradford Clay - 6. **Great Oolite, Stonesfield Slate, Fuller's Earth** - 7. **Inferior or Bath Oolite**. *c. Liassic Series.* 1. Alum shale 2. Marlstone 3. Lower Lias Shales.
		Lower Secondary or Triassic	1. **Upper New Red Sandstone**, Keuper or Variegated Marls 2. Muschelkalk (absent in England) 3. Variegated Sandstones, Bunter Sandstein, or Gres Bigarré.
	III. Palæozoic Epoch	Upper Palæozoic or Permian	*Upper Palæozoic,* or *Permian Series.* 1. Magnesian Limestone 2. Lower New Red Sandstone.
		Carboniferous	1. **Coal measures** 2. Millstone grit 3. **Carboniferous, or Mountain Limestone**
		Devonian or Old Red Sandstone	*Devonian, or Old Red Sandstone Series.*
		Silurian	*Silurian Series.* 1. Upper Silurian (Ludlow and Wenlock groups) 2. Lower Silurian (Caradoc Sandstone and Cambrian rocks)

Geological Column 2020s

Strata present in Geological Court (modern terminology)

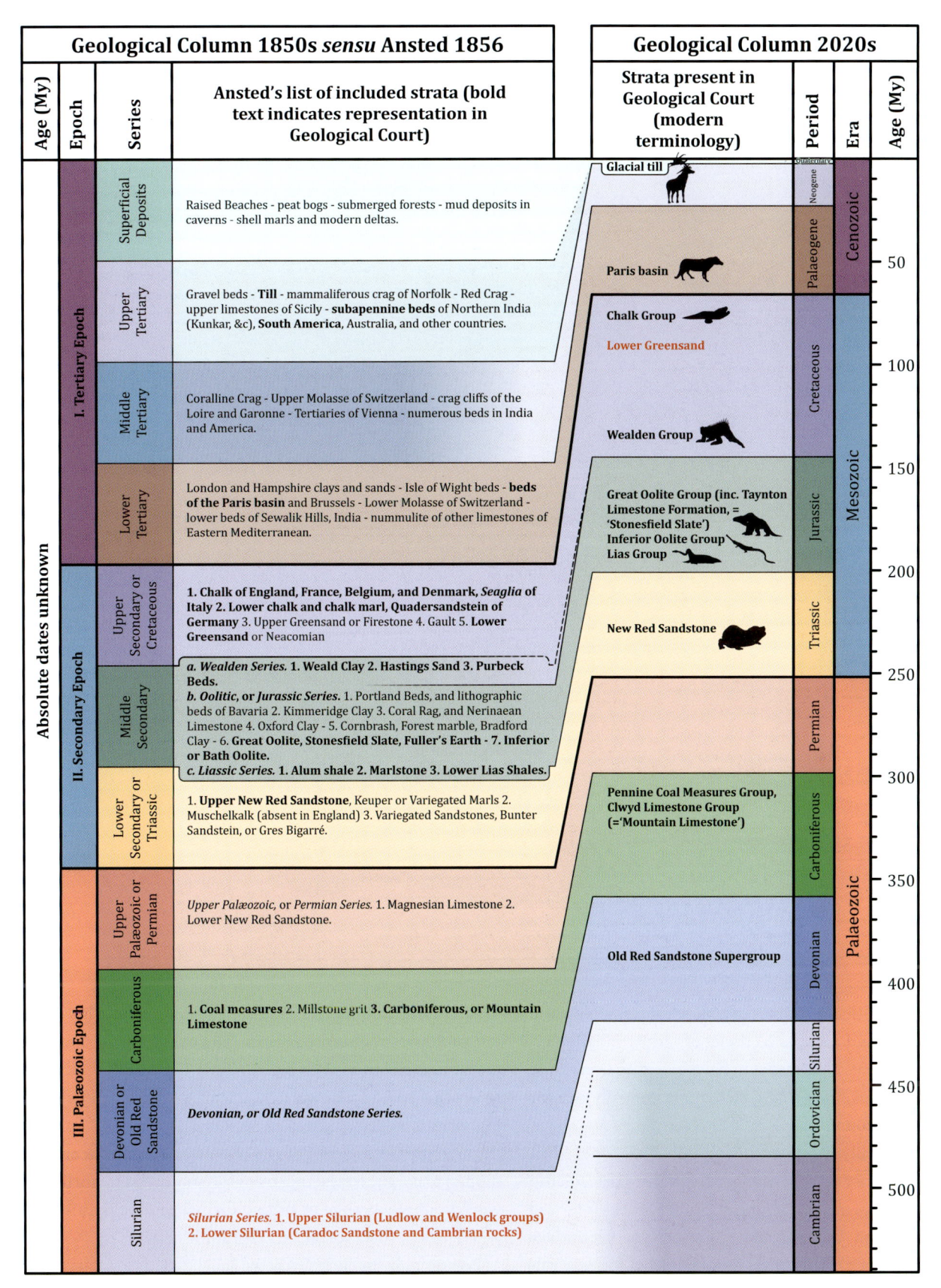

Strata	Period	Era	Age (My)
Glacial till	Quaternary	Cenozoic	
	Neogene		50
Paris basin	Palaeogene		
Chalk Group	Cretaceous	Mesozoic	100
Lower Greensand			150
Wealden Group			
Great Oolite Group (inc. Taynton Limestone Formation, = 'Stonesfield Slate') Inferior Oolite Group Lias Group	Jurassic		200
New Red Sandstone	Triassic		250
	Permian	Palaeozoic	300
Pennine Coal Measures Group, Clwyd Limestone Group (='Mountain Limestone')	Carboniferous		350
Old Red Sandstone Supergroup	Devonian		400
	Silurian		450
	Ordovician		
	Cambrian		500

Fig. 3.3 The Geological Column as understood by Ansted in the 1850s compared to that used in the modern day. With no robust methods of geological dating in the 1850s, the vintage column lacks a timescale. Details of the historic geological column are taken from Ansted (1856). Red text indicates strata planned for the Geological Court that was never implemented.

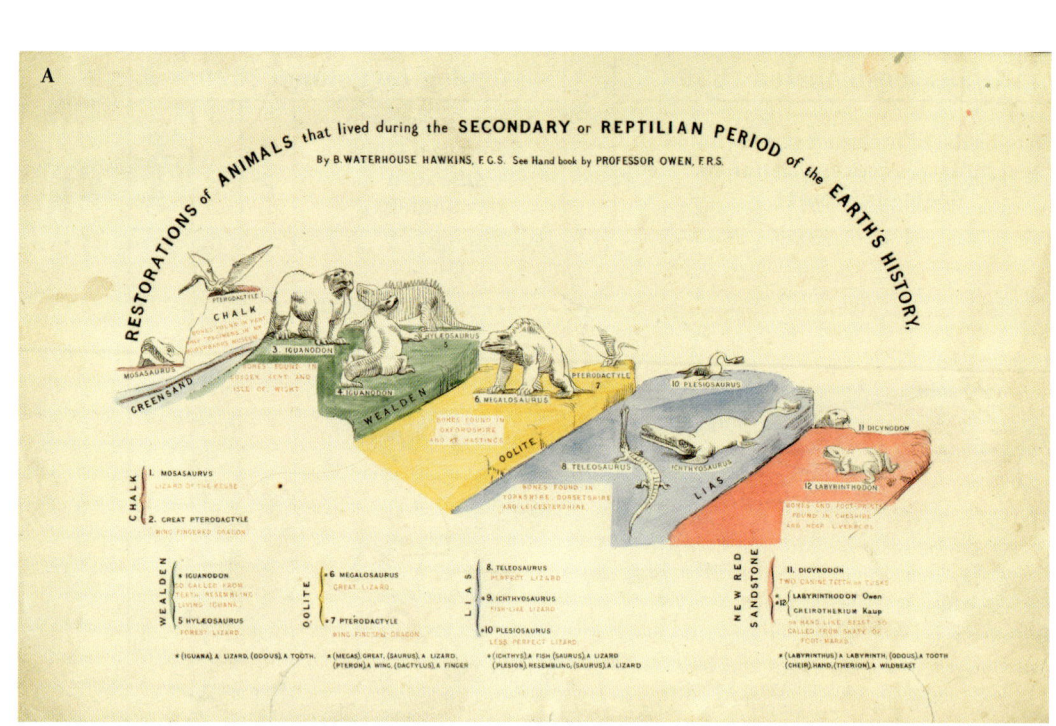

Fig. 3.4 The stratigraphic arrangement of the Geological Court as illustrated by Hawkins in context with the palaeontological displays. **A)** the archive illustration, *Restorations of Animals That Lived during the Secondary or Reptilian Period of the Earth's History* (date unknown); **B)** field-sketch style diagram of the same (from Hawkins, 1854).

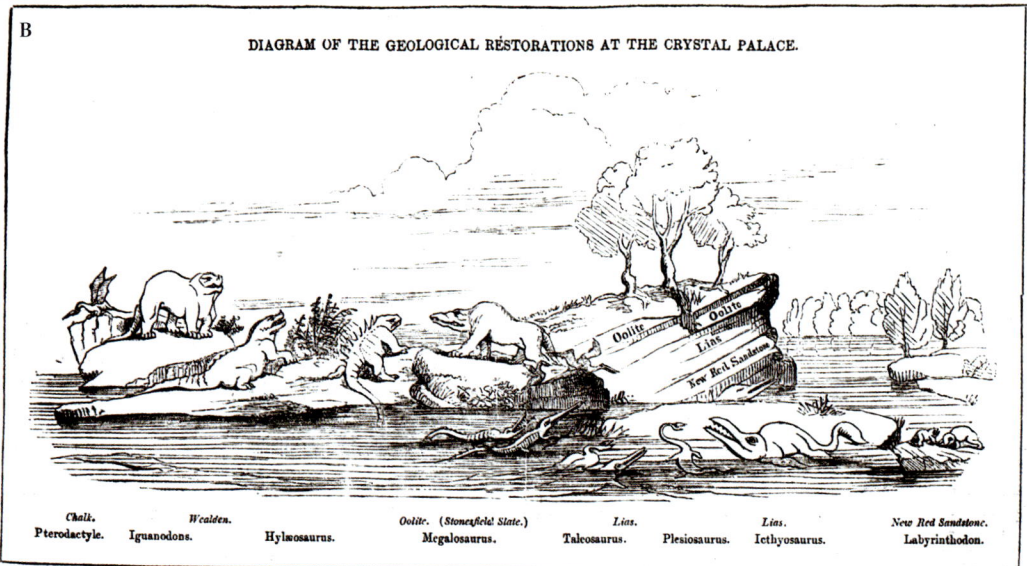

more than theoretical examples of geological principles, and instead introduced visitors to rocks of industrial and scientific importance. This seems to have been the overriding factor in decisions regarding which strata were chosen for the Geological Court.

Ansted generally arranged his chosen rock types conformably – in other words, younger rocks were placed atop older ones with no missing strata. Secondary rocks followed the Primary, and the Tertiary followed the Secondary (*see* Figs 3.1, 3.3, 3.4). This observation allowed for easy relative categorization of Earth's history and made it clear, for example, that dinosaurs existed more recently than the coal beds of the Carboniferous, and that the 'Age of Mammals' followed the time of great reptiles. To demonstrate this without building an enormous tower of rocks, Ansted imbued his simulated strata with a geological 'dip': a tilting of rock layers that spread their vertical stacking horizontally, allowing visitors to see changes in rock types simply by walking through the Geological Court. Echoing real fieldwork among dipped beds, guests walking from the southwest to northeast would sample a range of British formations well-known to geologists in their correct stratigraphic order.

Presumably for clarity, Ansted localized his Primary, Secondary and Tertiary rocks into largely segregated regions: a 'mainland' cluster of Primary strata, the Secondary Island (actually a large island with a few smaller islets in

44 Chapter 3 – Bricks, iron and tiles: rebuilding the past

Fig. 3.5 *Ichthyosaurus communis* and *'Plesiosaurus' macrocephalus* in the Tidal Lake. When the Crystal Palace first opened, the waters of this lake were raised and lowered by use of the park's mighty fountains, hiding and exposing the bottom regions of the sculptures. The exorbitant cost of moving so much water around reduced the 'tidal' nature of the lake to an increasingly infrequent occurrence, and by 1874 the feature was abandoned. (2017)

Fig. 3.6 Almost as spectacular as the Crystal Palace itself were the plentiful and gigantic fountains that occupied the park grounds, photographed here in 1854 by Phillip Henry Delamotte. Their operation demanded so much water be pumped up Sydenham Hill from the Tidal Lake that water levels visibly dropped, after which it returned via drains and pipes to rise again – a simulated 'tide'.

the southwest) and the Tertiary Island. This allowed the various waterways of the Geological Court to broadly mark the transitions between these epochs, although Ansted also provided geological continuity between all three regions in his displays. These simulated rock contacts provide observable boundaries between the Primary and Secondary and Secondary and Tertiary strata on dry land.

Ansted's arrangement wasn't without some intentional complications, however, which were introduced as a means to illustrate important geological principles and replicate regional geological complexity of UK strata (*see* Fig. 3.1). Faults and mineral veins run through his Primary strata and provide complex boundaries between these beds. Permian rocks, which would represent the latest portion of the Primary strata, are missing so that the Carboniferous Coal Measures contact the Triassic New Red Sandstone in a simulated unconformity: a gap in the geological record where sediments were eroded away before the deposition of overlying rocks. Ansted also introduced a series of faults between the region of the Primary and Secondary beds, representing major mechanical shifts in geological arrangements witnessed in nature as well as – on a more practical level – accommodating the expansive Secondary Island directly south to the Primary strata. This layout was explained to visitors in guidebooks (e.g. Phillips, 1856) but, with the Crystal Palace Company's strict policy about not using explanatory signage, these details were surely lost on visitors experiencing the park without such texts.

Ansted planned his Geological Court early in the development of the Crystal Palace Park and sought Paxton's approval before construction started. In addition to the building of artificial rock strata and palaeontological models, the Court was a major landscaping project with islands to be built and water courses to be dug. Paxton would have considered the cost and visual impact of the Court layout as much as the geological details (Doyle, 2008), in addition to pondering how Ansted's plan would have integrated with a grand and ambitious feature of the park grounds: an artificial 'tide' that raised and lowered the water level surrounding the Secondary Island. This system, designed in part by the famous engineer Isambard Kingdom Brunel, worked courtesy of the incredible and complex series of waterworks that fuelled the fountains of the Crystal Palace grounds – the largest of which could project water 50m into the air (Beaver, 1986).

The Tidal Lake of the Geological Court was plumbed into this system so that the water level rose and fell in concert with water movements elsewhere in the park. This feature was worked into the layout of the palaeontological sculptures where details of the basking marine reptiles were obscured by water at 'high tide' and only revealed when it was lowered. It is likely the Tidal Lake was not as regular as its natural counterpart, however, as the energy requirements – and thus expense – of pumping water to the Crystal Palace fountains was significant. At full operation over 30 million litres of water were moved around the park every hour – enough to fill over 400 12m-long shipping containers. The financial struggles of the Crystal Palace Company (Chapter 12) meant that use of the fountains diminished over time, likely ending entirely by 1874.

Having gained Paxton's approval with his plan, Ansted built a scale model of the entire Geological Court with colour-coded geological structures to guide construction (Philips, 1856). The Geological Illustrations were executed using a blend of conventional building materials – brick, cement and so on – and rocks imported to the park from elsewhere in Britain. A drive for authenticity meant that these rocks were sourced from the very deposits Ansted wanted to represent, and the scale of his displays made this no mean feat: hundreds of tonnes of rocks were transported from as far afield as Yorkshire and Bristol (*c.* 290km and 170km away from the park, respectively). This, presumably, is where James Campbell became heavily involved in the project, his role being to supervise not only the physical realization of the Geological Illustrations but also to source materials to build them.

Implementation of the Geological Illustrations varied in complexity and scope. At the simpler, often smaller-scale end of this spectrum were boulder, or boulder-sized bricks cemented onto pre-landscaped earth or stacked into small 'cliffs' and promontories. Others were enormous, complex structures comprising many tens of tonnes of rock integrated into the park's man-made topography and around Hawkins' sculptures. In such instances, brick structures overlain with flint pebbles provided the foundation for artificial cliffs and outcrops, over which the imported rock material was cemented to create realistic-looking geological features (*see* Fig. 3.2B; Doyle and Robinson, 1993). Various methods were used to achieve these effects, with some rocks cut into large blocks to create geological 'jointing' or facilitate construction of curving cliff faces, while others were attached as larger, naturally fractured pieces to form irregular rock faces.

Not all the rock types exhibited at Crystal Palace lent themselves to being transported from field locations – most notably clays and shales – and these were instead simulated using a cement render. The interior of the simulated lead mine, with its hanging stalactites, was also accomplished using cements over suspended metal frames. Photographs of the half-built Geological Court show that, where the Geological Illustrations were situated close to or under Hawkins' sculptures, the rock features were installed after the models had been secured in place on brick foundations. This, presumably, allowed for the palaeontological team to place their models without damaging the expensively-sourced overlying masonry.

Ansted's largest and most complex Geological Illustrations were situated in the northwest, 'Primary' region of the court, and they become increasingly simple towards the 'younger' strata of the Tertiary Island (*see* Fig. 3.1). The sophistication of the Primary sediments is probably related to the absence of palaeontological sculptures in this region of the park; without ancient reptiles or mammals to distract visitors, Ansted and Campbell could display their geology in full complexity and detail. The strata of the Tertiary Island are sparse and relatively simple, in part to reflect the less consolidated nature of Tertiary rocks, but also, perhaps, a result of the unfinished work of this part of the court. Sophisticated takes on Precambrian, Cambrian and Silurian strata – including displays of metamorphic rocks and crystalline basement rocks – were also planned (McDermott, 1854), but never executed.

To fully appreciate the detail of Ansted and Campbell's work, it is worth outlining the main features of their simulated geological landscape. These are discussed in further depth by Phillips (1854), McDermott (1854) and, more recently, by Doyle and Robinson (1993) and Doyle (2008). Though dilapidated, most of the Geological Illustrations remain *in situ* today and are worth your admiration when visiting the Geological Court. During such visits, however, be aware that the Geological Illustrations as seen today are a blend of original Ansted-Campbell creations and new elements constructed during major conservation work by the Morton Partnership in 2001 (*see* Chapter 13). This was conducted as authentically as possible from notes and direction provided by Peter Doyle, and the experience is thus broadly similar to that experienced by visitors in the 1850s.

Primary Strata

Exposed in the northwestern corner in the Geological Court, the Primary rocks are the most impressive Geological Illustrations of the display. Along with stepped, cliff-like 'outcrops' next to the Tidal Lake, a subvertical exposure was created adjacent to a narrow water body that, originally, was a spring feeding the wider lakes of the southern grounds. The latter is remarkably similar in appearance to an illustration in Ansted's 1856 *Geological Science* textbook (*see* Fig. 3.2A) and is the most elaborate set of Geological Illustrations, containing displays related to the industrial applications of geology and several important geological principles.

The oldest rocks of this sequence are those of the western extent and represent the Old Red Sandstone, a largely Devonian-age sandstone deposit representing a series of ancient rivers, lakes and dunes. An important deposit for

Fig. 3.7 'Outcrops' of Devonian Old Red Sandstone from the western banks of the Tidal Lake. **A)** a stepped cliff close to the path facing the *Iguanodon*; **B)** waterside boulders close to the weir adjacent to the Coal Measures. (2021)

early geosciences, the Old Red Sandstone crops out across southern and northern Britain, as well as in northern Europe and on the eastern seaboard of the United States. In the Geological Court, it also extends across the western shore of the Tidal Lake via simulated outcrops distributed around the water margins. The extent of these rocks are a clear indication that Ansted conceptualized a fault between the western 'mainland' and the strata of the Secondary Island (Doyle and Robinson, 1993), as there is a substantial gap in the rock record between the Old Red Sandstone and the Triassic rocks of the southwest Secondary Island (*see* Fig. 3.1). 18 tonnes of Old Red Sandstone blocks were obtained from the Bristol region for this display (Phillips, 1856).

Overlying the Old Red Sandstone was a complex construction representing the Carboniferous Mountain Limestone, a rock unit known today as Carboniferous Limestone. Perhaps the most extensive and ambitious single Geological Illustration, the Mountain Limestone was – and has since been rebuilt as – a large, cliff-like structure comprising 90 tonnes of rock sourced from Matlock, Derbyshire, arranged to resemble bedded strata overlying the neighbouring sandstone. Simulated joints and mineral veins break up the beds vertically, drawing the eye to a 27-tonne cap of Millstone Grit obtained from Yorkshire and Derbyshire. The lithified remains of a shallow, well-lit Carboniferous sea, the original Mountain Limestone was destroyed by alterations to the park watercourse in 1962 (Doyle and Robinson, 1993) and, by 2001, barely a trace of the original cliff remained. The structure that exists today is a historically accurate replica built after archaeological analysis found the original foundations, showing that Carboniferous Limestone blocks were arranged in a stepped fashion overlying an arched brick tunnel. The replica, as with Ansted and Campbell's original, was first planned with a 1:20 scale model, but used modern machinery to put the rocks in place. Architectural analysis shows that a pathway once ran directly alongside these cliffs, which has since been replaced with a more substantial watercourse.

Careful construction of the Mountain Limestone was especially required given that the rock face houses an entrance to another sophisticated Geological Illustration an artificial lead mine (*see* Fig. 3.8B–F). Also modelled on strata of the Matlock area, this mine descended into the earth behind the limestone cliff via a brick tunnel. McDermott (1854) stressed Campbell's role in the creation of this feature, stating that 'for nearly the whole of his life has been engaged in works connected with mines' (p. 202). The manufactured cave, formed to represent a lead-veined limestone cave, contained workers' tools as well as sculpted stalactites, stalagmites, flowstone and minerals to create 'everything essential to set up the working of a small mine under one's own account' (McDermott, 1854, p. 202). This feature was specifically included to demonstrate the economic importance of geology, showing visitors the conditions where lead – an important and versatile metal – was excavated. Investigations of the surviving mine in 2001 also revealed a hoofed statue foot: the remains of a lost *Megaloceros* statue that stood near the mine entrance (Chapter 4; Doyle and Robinson, 1993). Two access points – an entrance among the stepped limestone, and an exit behind and above the Primary Strata display – allowed visitors to pass through entirely. The mine was constructed as a sinuous tunnel that, if walking straight through, was about 20m long, but had an additional large chamber at the rear (*see* Fig. 3.8F). It was supplied with fresh air via vents positioned along its length and a limestone grotto, mimicking those naturally occurring in caves, was located close to the entrance.

Fig. 3.8 The recreated Carboniferous Mountain Limestone and cave. **A)** the cliff face as seen in 2013; **B)** cave entrance in 2021; **C–E)** scenes from the rear portion of the cave interior, showing the manufactured stalactites (C and E) and flowstone (D), 2021; **F)** map of the cave extent, based on 2001 work conducted by the Morton Partnership.

Much of the cave behind the limestone cliff has survived to modern times but the majority is inaccessible due to sediment accumulation and concerns over its structural integrity. Aspects of the mine entrance and immediate region of the adjoining tunnel were restored in 2001, but both entrances have been sealed with locked grills to deter vandalism and ensure public safety.

The grandest and most complex original Geological Illustration still in the Geological Court is located next to the Mountain Limestone: the Coal Measures (*see* Figs 3.2B–C, 3.9). These reproductions of Carboniferous beds are the only vintage part of the Geological Illustrations adjacent to the waterway that have survived intact to modern times. They once had a faulted relationship to the preceding limestone but this was destroyed by alterations to the watercourse adjacent to the Primary strata in the 1960s (Doyle and Robinson, 1993). Vertical sandstone blocks now stand on the site of this boundary. Unlike some Geological Illustrations, the Coal Measures are plainly visible when walking around the Geological Court and their faulted

Fig. 3.9 The Carboniferous Coal Measures, the largest and most complex original Geological Illustration left in the park, as seen in 2013. The waterfall and stream, flowing in this photo, operated until 2016.

nature is obvious, as is the juxtaposition of several lithologies: a sandstone base, a thick coal seam, an ironstone cap and bands of flint pebbles set into cement once covered in render (see Fig. 3.2B). The latter represented shale beds, rock types which had to be simulated because such lithologies would be too friable for transport to the park. The rest of the beds are genuine blocks of Coal Measure sourced from Clay Cross, Derbyshire, comprising 23 tonnes of coal, 4.5 tonnes of ironstone and 18 tonnes of sandstone (Phillips, 1856). Clay Cross also provided inspiration for the distinctive faulting and bed fabrics used in the display (Doyle and Robinson, 1993).

As with the Mountain Limestone, the Coal Measures demonstrated the economic importance of geology to visitors. The British Coal Measures were formed from the remains of Carboniferous swamps and forests and have been a source of fossil fuels for centuries, with particular significance to the Victorians. Coal was the fuel of the industrial revolution, a major factor in the success of the British Empire and thus, at a fundamental level, the very reason the Crystal Palace existed. Although this significance was overlooked by both Owen and Phillips in their official guidebooks, the importance of Ansted and Campbell's coal and ironstone representation was discussed at length in E. McDermott's (1854) *Routledge's Guide to the Crystal Palace and Park at Sydenham*:

> When the visitor leaves the Crystal Palace, he will be borne along by the steam-engine – the iron steed of the present day – which subsists on the primeval forests of this remote age, and he travels on iron rails, the metal of which was compelled to leave the unwilling ores by a resistless force, still inherent in those vegetable forms; he traverses the streets of the vast metropolis, or of the provincial town, illuminated by a gas obtained from vegetation, from which millions of ages have not yet sufficed to remove its carbon; arrived at home, he seats himself near his hearth, and while enjoying the warmth and comfort of his sea-coal fire, he knows that he is indebted for it to the forests of an extinguished world, – that the soot which forms on his chimney, or, it may be, falls in flakes upon his well-ordered room, is formed mainly of that carbon which was the principal element in those vast vegetable creations.
>
> MCDERMOTT, 1854, PP. 200–01

The series of Primary Strata terminates with a waterfall constructed using sandstone blocks. This section has been built in line with the rest of the section, using blocks imported into the park, but the origins of its rocks and their relationship to the surrounding 'strata' require investigation. No mention of a waterfall is found in original Crystal Palace guidebooks and the vertical alignment of the rocks is at odds with the carefully aligned dip of the other Geological Illustrations. These are good reasons to assume the waterfall is a modern addition to the Primary region, perhaps being installed as late as the 1960s.

Secondary Strata

Details about the Secondary Epoch Geological Illustrations are less forthcoming than those of the Primary strata because Crystal Palace guidebooks tended to sideline them in favour of discussing Hawkins' sculptures. They are mostly located on the Secondary Island but were connected to the Primary sequence via a 45-tonne band of Permian/Triassic New Red Sandstone, sourced from Bristol, which once overlayed the Coal Measures. As noted above, this association is a simulated unconformity that deliberately skipped rocks representing the Permian Period, but in turn established continuity between the Secondary and Primary strata. The New Red Sandstone is further seen in the southernmost regions of the Secondary Island, where a c. 2.5m-tall cliff exists alongside the '*Labyrinthodon*' and *Dicynodon*. A unique type of Geological Illustration – a fossil trackway, *Chirotherium* – was also once present here (*see* Chapter 10).

The New Red Sandstone represents the remains of a very hot, arid period in British history such as a desert or salt flat and crops out across much of the UK. The hard, well-consolidated nature of the New Red Sandstone has made it a popular rock for building, but this has not stopped the blocks adjoining the Primary strata crumbling since the 1850s, despite efforts to repair them in 2001. The Secondary Island New Red Sandstone 'cliff' is still relatively well represented, however (if very overgrown at time of writing) and, allowing for some restorative work in 2001, it is probably the most intact original geological feature after the Coal Measures.

Overlying the New Red Sandstone is the Jurassic Lias, represented by a sweeping limestone cliff set somewhat back from the marine reptile sculptures (*see* Fig. 3.10A). As their association with ichthyosaurs and plesiosaurs implies, the Lias limestones and mudstones represent remains of shallow seas. Several beds of limestone and dark rendering – simulating mudstones that would be impossible to transport to the site directly – were represented in this cliff, capturing details of the world-famous Lower Jurassic deposits of Dorset and Somerset (the former being widely known as 'The Jurassic Coast'). The original Lias cliff was dismantled and rebuilt in a guise faithful to the original in 2001.

Further along the island is another limestone 'outcrop', this time representing Jurassic Oolite: a remnant of lagoonal limestones from a tropical period in British history (*see* Fig. 3.10B–C). This sequence is divided across Lower and Upper Oolite, where the lower sequence is stacked into a (restored) 1.5m-tall cliff that once provided a perch for the now-missing Jurassic pterosaurs (*see* Fig. 3.10B). These rocks dip under the neighbouring Upper Oolite 'outcrop', a 1m-tall arrangement of stacked Portland Stone blocks that extend around the feet of the *Megalosaurus*. The entire Upper Oolite display had to be dismantled and rebuilt using original components in 2001 following chronic deterioration of the original. Two real Jurassic tree fossils, representing the Late Jurassic conifers *Protocupressinoxylon purbeckensis*, are situated along the sloping clifftop behind the *Megalosaurus* (*see* Fig. 3.10C). These silicified trunks were almost certainly sourced from the Purbeck Formation of Dorset or Swanage, UK. At one time large ammonites – real fossils, likely of the Jurassic genus *Titanites* – were present in the Tidal Lake, emerging into and receding from view as the waters rose and fell (*see* Fig. 3.2D; Doyle and Robinson, 1993). The effect may have recalled many palaeoartworks of the early twentieth century where invertebrate remains were scattered in the foreground of artworks depicting large extinct reptiles (e.g. Fig. 1.8B) but, today, only parts of these ammonites remain. They have suffered from crude restoration work in being repaired as circular instead of spiralled and also moved to the Wealden region of the island. Pieces of broken ammonite are still periodically recovered from the Tidal Lake.

Next in sequence are sandstone blocks representing the Early Cretaceous Wealden: a collection of sandstones and clays cropping out in the southeastern UK that represents ancient rivers and floodplains (*see* Fig. 3.10D–E). Of interest is that Ansted (1856) viewed the Wealden as a Late Jurassic deposit, rather than Cretaceous: one of the few major incongruences between his scheme and those of modern times (*see* Fig. 3.3). The Wealden rocks seen today are 2001 replacements of specimens which had, by the time of this conservation effort, all but disappeared from the park. Despite occurring alongside what is perhaps the most spectacular palaeontological region of the display – the two *Iguanodon* and *Hylaeosaurus* – the Wealden Geological Illustration is relatively modest: blocks of Hastings Sandstone arranged into a cliff at the feet of the standing *Iguanodon*, and a series of angled beds beneath the *Hylaeosaurus*.

This area of the Secondary Island was augmented, however, by artificial cycads of the genera *Cycas* and *Zamia* created by Hawkins and his team (Owen, 1854). A metal framework comprising a tripod of anchoring stakes and an internal skeleton supported a concrete cycad 'trunk' that held cast metal fronds via a metal ring at the tip. These are also no longer at the site as both the originals and later plastic replacements have been lost, probably to vandals.

Fig. 3.10 2001–03 restorations of the Geological Illustrations of the Secondary Island. **A)** the Lias Cliff; **B)** Lower Oolite cliff; **C)** Upper Oolite 'outcrop' with fossil tree trunks (*Protocupressinoxylon purbeckensis*); **D–E)** Wealden cliff under the feet of the standing *Iguanodon*, and bedding running towards *Hylaeosaurus*; **F–G)** Chalk cliffs – note the flint nodules in F. (2021)

Fig. 3.11 Manufactured Wealden cycad trunks, their fronds now missing, on the Secondary Island. (2018)

Sandstone blocks situated around the weir connecting the Secondary and Tertiary Islands represent a poorly documented Secondary Island Geological Illustration: the Lower Greensand. This collection of rocks from Late Cretaceous, shallow sea deposits cropping out throughout the southeastern UK were marked as present in the park on a sketch map by Hawkins (*see* Chapter 12) and have been observed in the Geological Court despite their omission from the Crystal Palace guides. Why they were not recorded in the 1850s is mysterious, but it may reflect the same circumstances that led to poor documentation of the Tertiary Strata (*see* below). The island sidewalls of both the Secondary and Tertiary Islands were constructed from Greensand boulders and some remain in situ today, but the bulk of the Greensand rocks were removed from the weir in 2016–17, curbing any further direct investigation into their presence and arrangement. It is likely, however, that the connection between the Tertiary strata and Greensand was used to create continuity between the Secondary and Tertiary Islands in much the same way that the New Red Sandstone connects the Primary and Secondary. Like the former, the Greensand/Tertiary boundary is an uncomfortable one, but it maintains the visage of a continuous rock outcrop as visitors progress from one epoch to the next.

The final stratum of the Secondary rocks is the Late Cretaceous Chalk, which was expressed as a 2.5m-tall limestone cliff serving as a platform for the large Cretaceous pterosaurs (*see* Fig. 3.10E–F). Chalk deposits represent ancient seas of several hundred metres water depth and famously form large cliffs and hills in various parts of the southern UK. Ansted and Campbell attempted to replicate these dramatic vistas using fitted Chalk blocks over a brickwork foundation, creating a structure large enough to be easily seen from the 'mainland' pathways. Although the brick foundations of this structure have survived to modern times, the overlying Chalk has not weathered well and has had to be reworked several times. At one stage the Chalk blocks were replaced with a different rock type, Portland Stone (Doyle and Robinson, 1993), but a Chalk face was reinstated on the cliff in 2001.

A smaller Chalk cliff – also heavily restored, to the extent of essentially being a new construction – exists adjacent to the taller, pterosaur anointed structure, but is often difficult to see owing to overgrown vegetation. Flint nodules, characteristic of those seen in Upper Chalk deposits of Britain, are evident at the base of the pterosaur cliff. A small amount of Chalk would have also been present on the southwest coast of the Tertiary Island had the Geological Illustrations been completed, adding to the continuity between Secondary and Tertiary strata afforded by the Greensand.

Tertiary Strata

The few Geological Illustrations of the Tertiary Island were little remarked upon by contemporary guidebooks, perhaps in part because the planned extent of this island was never realized. A sketch map produced by Hawkins in 1855 (*see* Chapter 12) shows the complex geological arrangement that was planned for this area (*see* Fig. 1.4), but there is no evidence that work on these strata ever started in earnest before the Geological Court project was stopped. The region housing the *Palaeotherium* and *Anoplotherium* has badly preserved remnants of a flint and concrete cliff likely representing aspects of the Palaeogene Paris Basin,

Fig. 3.12 'Tertiary' gravel beds created beneath *Megaloceros*, one of the few visible Geological Illustrations associated with the mammal sculptures. (2021)

reflecting the prevalent idea of nineteenth-century geology that more recent deposits formed in depressions of older rocks (McDermott, 1854). The *Megatherium* and *Megaloceros* stand on similar platforms. For the former, this represents glacial outwash found in the Argentine Pampas, and is almost always difficult to observe due to obscuring vegetation. The flint beds under the *Megaloceros* are formed of rounded London Basin chert pebbles sourced local to the Crystal Palace.

Since Ansted's day, both *Megatherium* and *Megaloceros* have shifted from the Tertiary to be classed as Quaternary genera (*see* Fig. 3.3), a division that – in part – reflects restructuring of the 'Tertiary Epoch' into the Cenozoic Era, and recasting of the Tertiary as an older and less expansive unit of geological time.

Constructing the palaeontological models

While Ansted and Campbell grappled with recreating a microcosm of British geology, Waterhouse Hawkins and his team were busy with the other half of the Geological Court: the thirty-seven life-sized paleontological sculptures. Although working towards the same goal, in most respects the two teams faced contrasting challenges and conditions. Where Ansted and Campbell were replicating modern natural spectacles, Hawkins and his team were imagining lost ones. While Ansted and Campbell had the full Geological Court to construct their displays, Hawkins and his team worked largely from a roughly built workshed. And while Ansted and Campbell could shape the Geological Court to their own liking, creating slopes or depressions to imbed and support their cliffs and escarpments, the palaeontological team had to engineer giant animal-shaped sculptures that could, for fear of otherwise ruining their impact, stand independently on their own four legs.

Sadly, we have no insight into this latter and most interesting aspect of Hawkins' work. No plans or prototypes are known and we cannot be certain how he, perhaps with ideas and assistance from others, developed the engineering approach used to build his sculptures. This is unfortunate as the constructional challenges facing the palaeontological team cannot be overstated. Today, models of giant prehistoric animals are constructed from lightweight, resilient materials such as foam, fibreglass and resin, so there's little doubt that they will be able to support their own weight with appropriate engineering care. Hawkins, however, had to construct his models from traditional, heavy building materials: bricks, tiles, cement and iron. To minimize costs some of these were recycled from other building projects: specifically, the deconstructed medieval building Gerard's Hall undercroft (McCarthy & Gilbert, 1994). Holding the models together was Portland cement (identified in the palaeontological structures by M. Eden, Sandberg LLP and R. Siddall, University College London) which, to date, had yet to be applied to a non-architectural project. Despite using unconventional materials, Hawkins set the bar high for reconstruction quality, refusing to add any obvious or unnatural bracing to the giant models, feeling that:

Fig. 3.13 A c. 1854 photograph of the Secondary Island under construction. Both *Iguanodon* appear finished, but scaffolding and tents cover the in-progress *Megalosaurus* and *Hylaeosaurus*. *Teleosaurus* and fragments of a large ichthyosaur, perhaps *Ichthyosaurus communis*, can be seen to the right of the *Megalosaurus* tent. Photograph probably by Philip Henry Delamotte. *See* Figs 3.15–16 for views of the same site at different angles.

Fig. 3.14 The Secondary Island in 1855, showing completed *Iguanodon*, *Hylaeosaurus* and '*Pterodactylus*' *cuvieri*. Blocks of Chalk lay behind the dinosaurs to assemble the Geological Illustration below the pterosaurs. Photograph probably by Philip Henry Delamotte, and probably part of the same series as Fig. 3.17.

Fig. 3.15 Marine reptiles and a curiously placed mammal on the nascent banks of the Secondary Island, c. 1854. *Plesiosaurus dolichodeirus* and '*Plesiosaurus*' *hawkinsii* rest on wooden sleds next to the fully installed '*Ichthyosaurus*' *platyodon*. A single '*Labyrinthodon pachygnathus*' can be seen in the far right, and *Palaeotherium magnum* rests on its side in the bottom left. Plaster jackets with skin impressions are cast around the right of the image and stone blocks in the background may represent material for the Lias cliff. Ghostly figures – photographic artefacts – capture Geological Court workers moving around the site in the distance. Photograph probably by Philip Henry Delamotte (*see* Figs 3.13 and 3.16 for views of the same site at different angles).

Fig. 3.16 Northeast view of the Secondary Island, c. 1854. '*Labyrinthodon*' and *Dicynodon* (barely seen) are in the foreground, with some of the former surrounded by pieces of plaster. Enaliosaurs are seen to the left. *Plesiosaurus dolichodeirus* and '*Ichthyosaurus*' *tenuirostris* are not yet fixed in place, with the latter yet to be moved to its final position. An armless *Teleosaurus* with a propped-up jaw faces two cliff-less '*Pterodactylus bucklandi*', behind which seems to lie pieces of *Ichthyosaurus communis*. Dinosaur-covering tents and part of the completed *Iguanodon* are seen in the distance. Photograph probably by Philip Henry Delamotte (*see* Figs 3.13 and 3.15 for views of the same site at different angles).

Fig. 3.17 A rarely seen, grainy 1855 photograph of the Tertiary Island during construction in late spring or summer 1855, likely by Philip Henry Delamotte. *Palaeotherium magnum* and '*Pa. minus*' can be seen, as can the head of one *Anoplotherium*. In the background, construction of the Chalk cliff around the large pterosaurs is underway, while work appears to be completed on the standing *Iguanodon* and *Hylaeosaurus*. This photo is probably complementary to Fig. 3.14.

…their natural history characteristics would not allow of my having recourse to any of the expedients for support allowed to sculptors in an ordinary case. I could have no trees, nor rocks, nor foliage to support these great bodies, which to be natural, must be built fairly on their four legs.

HAWKINS, 1854, P. 447

With a gargantuan task and little money to waste, the construction of Hawkins' models had to be meticulously planned and executed with great care over their thirty-two-month build time. Although we lack the full picture of how the Crystal Palace models were made, a number of sources – accounts from Owen (1854) and Hawkins (1854), press articles, the surviving models, illustrations of Hawkins' workshop, and photographs of the models being built and installed on the Geological Court – permit a reasonable insight into their creation. Especially good data are provided by photographs likely taken by Philip Henry Delamotte, who documented the entire development of the Crystal Palace site (Delamotte's work is synonymous with the Crystal Palace – *see* Leith 2005 for a compilation of his shots). Several authors have used these sources, as well as archived correspondence, to compile accounts into the development of the models (Doyle and Robinson, 1993; McCarthy and Gilbert, 1994; Secord, 2004; Doyle, 2008; Dawson, 2016), which we have summarized and augmented here.

Work commenced on the sculptures in September 1852. Before any three-dimensional work began, Hawkins produced preliminary sketches of each species to establish their proportions, poses and appearance, but no examples of such works seem to have made their way into archives. Comparable sketches and rough drafts survive for his other palaeoartworks, however, showing us his creative process and attempts to work out compositions, poses and anatomy. From these and other drawings, Hawkins then created 6th- or 12th-scale models in clay which were shown to Owen for comment and revised as necessary (this may be the only time Owen had any involvement in their development – Secord, 2004).

Unfortunately none of these models are definitively known today, but a series of clay and plaster sculptures archived in the Wisbech & Fenland Museum, Cambridgeshire, are suspected to have played some role in the early development of the Geological Court. This collection of twenty-five models includes most of the Crystal Palace species, excepting the pterosaurs, *Mosasaurus*, *Anoplotherium* and *Xiphodon*, as well as some planned, but never built, species: a mammoth and a giant tortoise (likely '*Colossochelys*'). Mike Howgate (2015) speculated that these were used to plan the layout of the Geological Court in three dimensions, as well as serving as possible references for the famous 1854 Geological Court print by George Baxter (*see* Fig. 2.5). The creatures in Baxter's print – which was executed before Crystal Palace

Chapter 3 – Bricks, iron and tiles: rebuilding the past **55**

Fig. 3.18 Archived draft artworks of extinct animals (pterosaurs, plesiosaurs, mammals) by Waterhouse Hawkins. No plans or drafts of the Crystal Palace dinosaurs are known but drawings of this nature must have been made to conceptualize the palaeontological sculptures.

was opened, and contains several park elements that were replaced or modified before the opening day – certainly resemble the Wisbech models, potentially providing an answer to how Baxter knew of the sculptures' form before Hawkins finished their construction (Cain, 2014).

With preliminary sketches and scale models approved (or at least seen) by Owen, production on the full-size sculptures could begin. To accommodate the construction of thirty-seven life-sized prehistoric animals a dedicated workshop was built on the Crystal Palace grounds. With money being relatively tight, however, Hawkins' workshop was not a luxurious studio. Described as a 'long, low, window-roofed building' (Anonymous, 1854, p. 280), it was a cramped and draughty shed that was almost inaccessible in bad weather: 'a rude and temporary wooden building almost inaccessible for deep ruts and acres of swamp and mud – a miniature saurian bog' (McDermott, 1854). Terrible conditions plagued some of the build, such that Hawkins and his team had to protect themselves 'from winter wet and blast' with mud-boots and warm clothing (Dawson, 2016). One illustration of the workshop even shows rats and birds busying themselves on the floor (*see* Fig 1.12). But despite the wretched working conditions, visitors were astonished by what the workshed contained:

> Never before has there been gathered under a single roof such an array of the wonders of geology. One might well be excused a momentary tremor of alarm, did not the chiselling and modelling of the dusty-frocked artists assure us that the grim monsters around were but stone or clay. Covering the floor, towering to the roof, and crowded into every corner, are huge lizards, and turtles, and long-snouted crocodiles, and hideous reptiles of fish-like, frog-like, bird-like

56 Chapter 3 – Bricks, iron and tiles: rebuilding the past

Fig. 3.19 The small Crystal Palace dinosaur models archived at the Wisbech & Fenland Museum, Cambridgeshire. These may have played some role in the planning of the Geological Court, but their exact significance is unknown. Of note is that the set of twenty-five models also includes species planned, but never finished, for the display. **A)** *Ichthyosaurus communis*; **B)** *Iguanodon*; **C)** *Hylaeosaurus*; **D)** *Dicynodon*; **E)** *Megaloceros*; **F)** *Mammuthus* – presumably a plan or study of the woolly mammoth intended for the Tertiary Island.

forms, with great shaggy-haired beasts of strange aspect, all seeming to snarl and glare upon you, as you move about amongst them.

ANONYMOUS, 1854, P. 280

Several images of the workshed show such conditions, the building packed with models in various states of completion (*see* Figs 1.12, 3.20–21). The wonders of Hawkins' workshop attracted esteemed visitors, including Queen Victoria (Secord, 2004) and numerous journalists. Guests would have seen that, while not the sole artist on this project, the final details of the models – skin texture, musculature – were added by Hawkins himself (Hawkins, 1854).

Manufacturing the models was a complex task. Because the subject species differed so much in size and proportion, no single method of construction was suited to them all. Instead, several different approaches were used depending on the size and anatomical complexity of each animal as well as – of course – the budgetary constraints of the project (Doyle, 2008; Brierley *et al.*, 2018). Most, perhaps all species were first realized in clay at full size before they were rebuilt with bricks, iron, concrete and tiles.

Fig. 3.20 Images of the palaeontological workshed featured in the press. **A)** *Iguanodon*, *Megaloceros*, '*Labyrinthodon*' and *Megatherium* alongside three figures. Note the wheeled cart under the *Megaloceros*; **B)** *Megatherium* is chiselled from limestone; **C)** one of the most informative images of Hawkins' workshed: the clay *Iguanodon* being cast in plaster to transfer details of its skin to the iron and concrete version being built in the Palace grounds, alongside a scale maquette and boards of sketches and designs. *Megaloceros* and *Palaeotherium*(?) lurk in the background. A) and B) from *Die Gartenlaube* (1854 and 1853, respectively); C) from *The Official Illustrated Guide to the Brighton and South Coast Railways and Their Branches* (1855).

Chapter 3 – Bricks, iron and tiles: rebuilding the past

Fig. 3.21 Further press images of Hawkins' workshed. **A)** a packed room featuring, clockwise from top left, *Megatherium*, the snout of a '*Labyrinthodon*' (probably '*L. pachygnathus*'), a barely-seen *Palaeotherium magnum*, the clay standing *Iguanodon*, *Mosasaurus*, *Megaloceros*, *Ichthyosaurus tenuirostris*(?), the snout of *Teleosaurus* and *Plesiosaurus dolichodeirus*. The *Megaloceros* base reads 'B. Waterhouse Hawkins *fecit*', identifying Hawkins as the artist behind the models; **B)** another view of models in storage. From left to right, '*Labyrinthodon pachygnathus*', *Palaeotherium magnum*; the four '*Anoplotherium gracile*'; *Megaloceros* stag and doe, and *Dicynodon*.

With the exception of the dinosaurs, it is difficult to know which models in the workshed illustrations are clay moulds and which are finished models. The clay sculptures captured, at full scale, details of skin and facial anatomy that could be cast in plaster and then impressed into the drying fine-sand concrete skin of the brick and iron models. Because the clay and concrete models had to be identical for this construction method to work, it was critical that they followed the same plans in exacting detail. One workshed sketch shows that the creation of the clay versions was guided by scale models and artwork (*see* Fig. 3.20c), and correspondingly detailed plans were presumably drawn up for their concrete and brick counterparts (*see* Bramwell and Peck (2008) for examples of Hawkins' intricate construction plans from other projects). Illustrations and photographs show that smaller models were constructed on wooden platforms of varying kinds, which likely helped the construction team move them around the workshed and beyond. The models were packed tightly alongside each other as construction progressed, sometimes being stored in racks to clear floor space.

There is good evidence that at least the smaller completed models were not, as has sometimes been suggested (Doyle, 2008), built in situ on the Geological Court. Rather, contemporary photographs and news articles reveal how they were moved to their locations on sledges and wheeled carts, having been built entirely in the worked first. We can see in one photograph, for instance, the seemingly completed *Plesiosaurus dolichodeirus* on the banks of the proto-Secondary Island resting on a wooden sled identical to those featured in an 1853 illustration of Hawkins' workshop (*see* Figs 3.14, 3.21A). Other photographs show models – plesiosaurs, *'Ichthyosaurus' tenuirostris* and *Palaeotherium magnum* – out of position, resting on slopes or located several metres from their eventual locations (*see* Figs 3.15-16). In one photo, *Pa. magnum* (barely in shot, and identified only by its feet and trunk) is lying on its side (*see* Fig. 3.15), perhaps resting on an out-of-shot sled or cart. An April 1855 *Observer* article describes the issues of moving even these small models around, detailing the transfer of a *'Pterodactylus' cuvieri* model from Hawkins' worked to its perch atop the Chalk cliff:

> The conveyance of this creature from the *atelier* of Mr. Hawkins, in which he was created… was attended with considerable difficulties. His weight considerably exceeds one ton, and the physical construction of the animal afforded but few facilities for locomotion…
> A strong raft was accordingly constructed, powerful shears erected on the edge of the lake and, after much difficulty, and with many false starts, the animal was embarked and floated successfully over the waters of the lake, accompanied by Mr. Hawkins, who evidently regarded his interesting *protégé* with all the fond affection of a parent, his arms being tenderly thrown around the neck of his child…
>
> THE OBSERVER, 15 APRIL 1855, P. 4

Pieces of plaster cast, some with obvious skin impressions, can be seen surrounding the models in Delamotte's photographs, perhaps having been used to protect the models as they were transported around the site. Once situated in place, most of the models were concreted into place and are now entirely immobile. They can, of course, be worked free from these bases to regain a degree of transportability: this has been achieved for the smaller mammal statues, which were relocated several times in the late twentieth century (Doyle and Robinson, 1993).

Other approaches were used for larger sculptures. In the background of one Delamotte photograph are (seemingly hollow) components of what appears to be an ichthyosaur model (probably *'Ichthyosaurus' communis*), including one piece with an unmistakable eye (*see* Fig. 3.14). The same photo shows the majority of a *Teleosaurus* without its right forearm attached. This suggests such models were built in sections in the worked and then assembled on the Geological Court grounds, a more manageable approach than moving enormous, fragile and heavy brick and concrete sculptures in one piece. This method may have typified construction of many of the 'mid-sized' models and, as with their smaller brethren, they were set in concrete foundations once in place.

Some smaller species presented technical conundrums that exceeded sculpture mobility. While most of the reptile species were low-bodied, sprawling and thus easily stabilized, the slender-limbed, top-heavy mammals and pterosaurs presented greater structural challenges. The limbs of such constructions would be prone to breaking if made entirely from brittle material like concrete, so they were instead built around iron skeletons that anchored them in place with some cast-metal 'skin' around their internal frames (Doyle, 2008). These iron supports are well hidden in the mammal models but can be seen upon close inspection of the pterosaur wings where damage has rendered parts of the framework visible.

Iron bracing likely served a particularly important purpose for the *Megaloceros*, strengthening the neck and head against the weight of real fossil antlers. The use of real fossils in an outdoor display may seem strange, but it was a canny decision by the palaeontological build team. *Megaloceros* antlers are found in their hundreds in some parts of the UK so were not only relatively available, but – on account of their relatively recent geological age – are still surprisingly strong structures capable of supporting their own weight even when only anchored at their burrs. Hawkins and his team might have been able to manufacture antlers using iron frameworks as they did the pterosaur wings, but they would likely have been structurally weaker and more prone to weathering than real specimens. The rearing *Megatherium*, which sits on a robust base of crouched legs with a supportive tail, did not need a metal skeleton and was instead, seemingly uniquely, carved from blocks of Bath Stone – a type of limestone (*see* Fig. 3.20B; Doyle, 2008).

Of course, as the largest sculptures, the dinosaurs presented the greatest difficulties for Hawkins and his team.

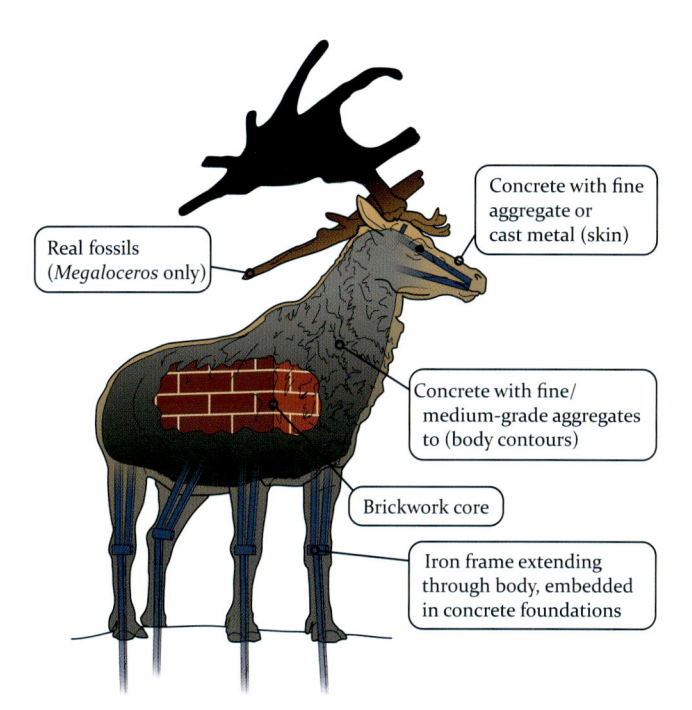

Fig. 3.22 Simplified diagram of presumed construction methods used in smaller palaeontological sculptures, based on details observed in conservation photographs.

Labels (Fig. 3.22):
- Real fossils (*Megaloceros* only)
- Concrete with fine aggregate or cast metal (skin)
- Concrete with fine/medium-grade aggregates to (body contours)
- Brickwork core
- Iron frame extending through body, embedded in concrete foundations

There was simply no way to move these giant models from the workshop once they were built: even the clay versions weighed in the region of 30 tonnes (Hawkins, 1854) and required structural reinforcement to stay upright (visible in Figs 1.12. 3.20C, 3.21A). Hawkins described their construction as 'not less than building a house on four columns' (Hawkins, 1854, p. 4) and, like houses, each model had deep foundations – built from bricks beneath their feet and tails – and an internal frame – thick iron bars (Doyle, 2008). Digging these foundations meant that completion of the dinosaur sculptures took priority over shaping the topography of the Secondary Island, which would have been impossible to finalize before the enormous dinosaurs were finished. Delamotte's photographs accordingly show the bizarre site of a crudely realized, unvegetated Geological Court with gleaming, fully detailed dinosaurs perched on muddy slopes. These same photos show covered scaffolds and tents erected around the dinosaur construction sites (specifically, *Hylaeosaurus* and *Megalosaurus*), presumably to accommodate working on the half-finished models during bad weather.

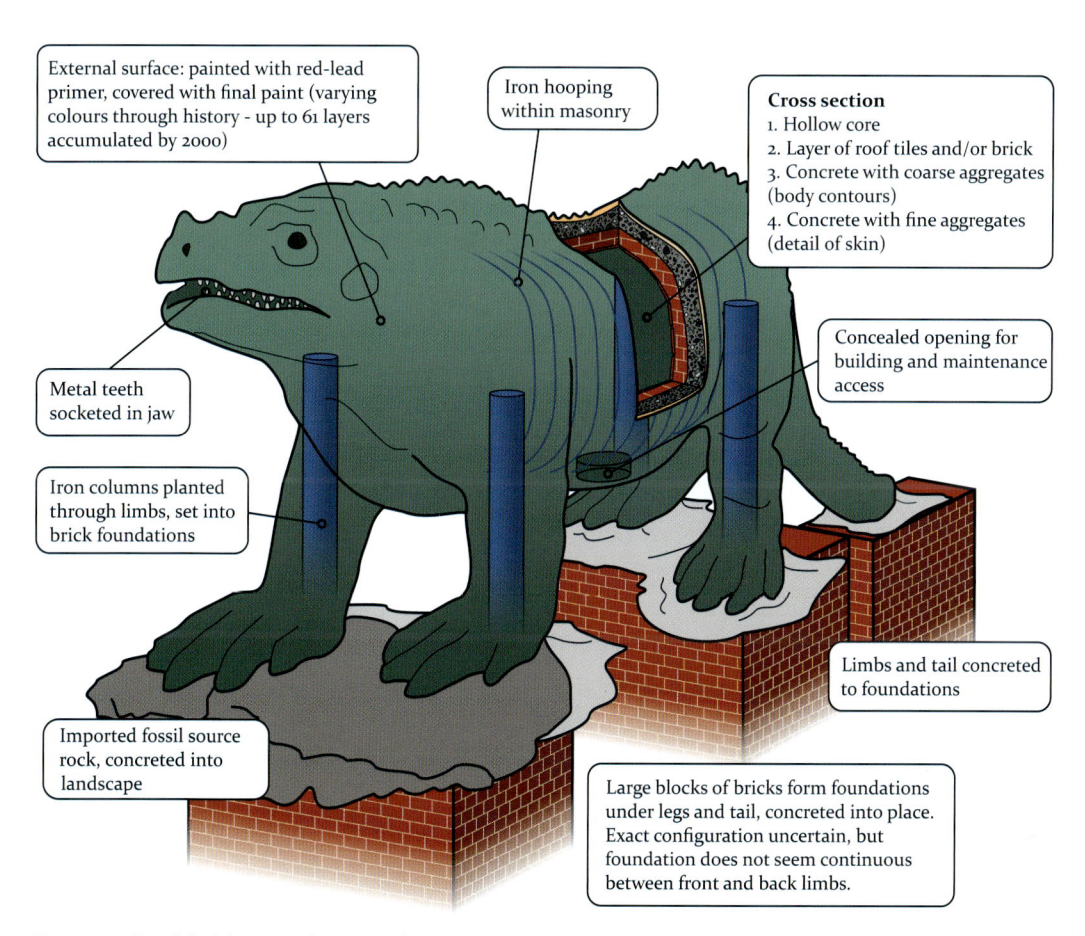

Labels (Fig. 3.23):
- External surface: painted with red-lead primer, covered with final paint (varying colours through history - up to 61 layers accumulated by 2000)
- Iron hooping within masonry
- **Cross section**
  1. Hollow core
  2. Layer of roof tiles and/or brick
  3. Concrete with coarse aggregates (body contours)
  4. Concrete with fine aggregates (detail of skin)
- Concealed opening for building and maintenance access
- Metal teeth socketed in jaw
- Iron columns planted through limbs, set into brick foundations
- Imported fossil source rock, concreted into landscape
- Limbs and tail concreted to foundations
- Large blocks of bricks form foundations under legs and tail, concreted into place. Exact configuration uncertain, but foundation does not seem continuous between front and back limbs.

Fig. 3.23 Simplified diagram of presumed construction methods used in the dinosaur sculptures, based on details given in Hawkins (1854). Some details, such as the extent of the brick foundations, are uncertain.

Chapter 3 – Bricks, iron and tiles: rebuilding the past   61

Despite having hollow bodies to minimize their weight, the material inventory for each dinosaur is impressive. Hawkins (1854) listed the material for the standing *Iguanodon* as four iron columns (each nine feet long, seven inches wide), 600 bricks, 650 five-inch half-round drain tiles, 900 plain tiles, 38 casks of cement, 90 casks of broken stone, 100 feet (30.5m) of iron hooping and 20 feet of cube inch bar. Recent studies of the models have shown how these elements were combined to make the dinosaur sculptures (*see* Fig. 3.23; Brierley *et al.*, 2018).

Once the foundations and iron columns were in place, Hawkins and his team created the core of his models using bricks, tiles and iron hooping, over which a layer of coarse concrete was applied. This concrete transformed the basic shape of the underlying brick and tiles into a more animalistic shape, but was too coarse to capture the anatomical details of the skin and face. Such finer elements were instead achieved with the same sandy concrete that acts as the 'skin' of virtually all the models. Cracks and joins in the cast skin panels were sealed with mortar, and the *Iguanodon* and *Megalosaurus* had cast metal teeth drilled into their open mouths. The standing dinosaur models were left with concealed openings in their bellies wide enough to climb into for building and maintenance access, as well as to facilitate rainwater escape. It's within the models that we find a uniquely personal record of their construction: carved into the lower jaw of the standing *Iguanodon* is Hawkins' own signature – 'B. Hawkins, Builder, 1854'.

Having constructed the models, a layer of protective red lead oil paint was first added, before a secondary coat of white lead paint was applied to prepare the models for their intended colouration. Unfortunately – and as will become apparent in subsequent chapters – documentation of the models' colours is very poor and our records only begin in earnest in the mid-twentieth century when park visitors began taking colour photographs and video. Relying on media archives or photos taken by the public biases our colour information to the more popular and obviously seen models, so we have only patchy data on the hues sported by the less famous sculptures even in recent decades. An attempt to deduce their original colours was made in 2000 when Hirst Conservation analysed paint layers from the models to identify different colour layers. Although evidence was found that the original paint had failed within a few decades, traces of the original hues were still detectable – sometimes under sixty-one additional coats of paint (Hirst Conservation, 2000) – and it can be ascertained that many models were first painted earthy greens, greys and browns.

Unfortunately, the paint on sculptures situated in water was often too decayed for such analysis, the presence of moisture accelerating their decay. Evidently the paints used by Hawkins and his team were inexpensive and not long-lasting: a likely consequence of the economical manner in which the models were built.

Hawkins' workshed in the press: exploiting dinomania
With the Geological Court part of an enormous commercial enterprise, the advertising value of the palaeontological sculptures was seized upon by the Crystal Palace Company. By January 1854 the *Morning Chronicle* reported some 'ten or fifteen of the sculptures had been completed, providing a foundation upon which the Company could start to build hype and anticipation for their novel prehistoric enterprise and, by extension, the wider park project they were attached to. Among other activities, journalists and draftsmen were given access to the workshed in late 1853 and early 1854 to produce magazine and newspaper articles on the developing Geological Court project.

Such articles appeared in periodicals and newspapers across Europe, often with lavish illustrations of the creatures stored in Hawkins' workshed (*see* Figs 1.12, 3.20–21). The appeal of the palaeontological sculptures was strong, such that even articles providing overviews of the entire Crystal Palace project would feature images of the prehistoric animals instead of images of the Palace, Park grounds or other exhibitions. These decisions to showcase Hawkins' models over other, sometimes equally fantastic, Crystal Palace subjects represent early recognition of the marketing power of charismatic fossil species.

Accounts of the Geological Court preparations appeared in several widely read venues, including the *Illustrated London News*, the *Morning Chronicle*, *Punch*, *Hogg's Instructor* and, in France and Germany, *L'Illustration* and *Die Gartenlaube*. Although some coverage was brief, many reports spanned multiple pages and provided detailed insights into the construction of the Geological Court. Such lengthy pieces were required to not only generate excitement around the Geological Court but to convince a public unaccustomed to the appearance of extinct animals that the products of Hawkins' workshed were not exercises in fantasy. Journalists became educators by introducing the concepts of fossils, geological time and extinct animals, stressing that the species displayed at Crystal Palace were real, albeit now extinct animals and not imagined monsters.

Dinosaurs, especially *Iguanodon*, were the focus of many articles. These species were, as the largest and most spectacular denizens of the Court, both the biggest draw but also the hardest sell, in terms of scientific validity, for the Crystal Palace Company. As expressed by an anonymous author in *Hogg's Instructor*:

> ...there are thousands of even well-educated persons who will be altogether unprepared for this novel and startling display. In many cases, no doubt, the first impression will be that the monstrous and hideous creatures represented are merely fanciful and imaginary, as little like any animals that have actually lived upon the earth as Mr Layard's winged bulls and eagle-headed monsters from Nineveh, or the sphinxes and colossal gods of Egypt. But only in the case of the very ignorant will that impression last. To all who can and care to read, a threepenny guide-book, from the pen of Professor Owen himself, will disclose the true significance of the spectacle before them, and reveal, in the strange array of animal forms exhibited, the predecessors of our race in the occupation of the earth, the forms of the various beings that had life upon it during the successive epochs in which it was being fitted and prepared for the abode of man.
>
> ANONYMOUS, 1854, P. 286

The grandest public relations event concerning the Crystal Palace dinosaurs is also the most famous: the 1853 New Year's Eve banquet that took place inside the clay mould of the standing *Iguanodon*. This seven-course meal was not, as is sometimes asssumed, to celebrate completing work on the sculptures (on the contrary, work on them would continue for another seventeen months), but was an opportunistic publicity stunt to showcase the intellectual and artistic work commencing in the Geological Court and impress journalists and newspaper editors. Hawkins, the host of the meal, may have been inspired by a similarly eccentric 1802 dinner held within a mastodon skeleton that had been attended by his father (McCarthy and Gilbert, 1994) or more contemporary banquets held within grand construction projects (e.g. Brunel's 1843 Thames Tunnel or Wyatt's 1846 bronze statue of the Duke of Wellington on his horse).

Invitations were sent with only one week's notice (McCarthy and Gilbert, 1994), but among the attendees were high-profile scientists, newspaper moguls, and senior members of the Crystal Palace Company board. The guest of honour was Richard Owen, who attended for one of his only documented visits to Hawkins' workshop (Dawson, 2016).

The *Iguanodon* banquet is an oft-discussed event in the history of palaeontology, and not just for its bizarre spectacle: it also has bearing on the emerging popularization of dinosaurs and palaeontology, as well as Owen's rising reputation in scientific and political culture. Much is known about the banquet thanks to contemporary reports, illustrations drawn at and after the event, copies of invitations and menus, as well as an 1872 write-up by Hawkins himself (McDermott, 1854; Owen, 1894; Rudwick, 1992; McCarthy and Gilbert, 1994; Cadbury, 2000; Secord, 2004; Bramwell and Peck, 2008; Dawson, 2016). Hawkins hosted his banquet in the clay mould of his standing *Iguanodon* on account of it being not only large, but among the earliest models to be finished. In a later recollection of the event, he wrote:

> The restoration of the *Iguanodon* was one of the largest and earliest completed of Mr Waterhouse Hawkins gigantic models, measuring thirty feet from the nose to the end of the tail, of that quantity the body with the neck contained about fifteen feet which when the pieces of the mould that formed the ridge of the back were removed the body presented the appearance of a wide open boat with an enclosed arch seven feet high at both ends.
>
> HAWKINS, 1872
> (QUOTED IN BRAMWELL AND PECK, 2008, P. 25)

A marquee or tent, pink and white in colour (McDermott, 1854), was erected or suspended around the model. The exact location of the dinner is not recorded, but it must have taken place in Hawkins' workshop. Details of certain banquet illustrations, including some of Hawkins' draft invitation designs, match aspects of the workshed seen in other artwork, such as the wooden roof, raised platforms around the model, and plaster casts on and around the *Iguanodon* mould. A *Morning Chronicle* write-up also notes the close proximity of other models – including a *Plesiosaurus* – which does not match the layout of the Geological Court. The challenges of moving a 30-tonne clay model, the relatively late notice of the event, and the practical demands of hosting a banquet in the middle of winter, also point to the workshed as being the likely venue.

In this context, the pink and white marquee probably added a festive quality to the workshop as well as providing additional protection from the cold winds and draughts said to have characterized the evening (McCarthy and Gilbert, 1994; Secord, 2004). Above the sculpture, a chandelier and banners naming prominent figures of nineteenth-century

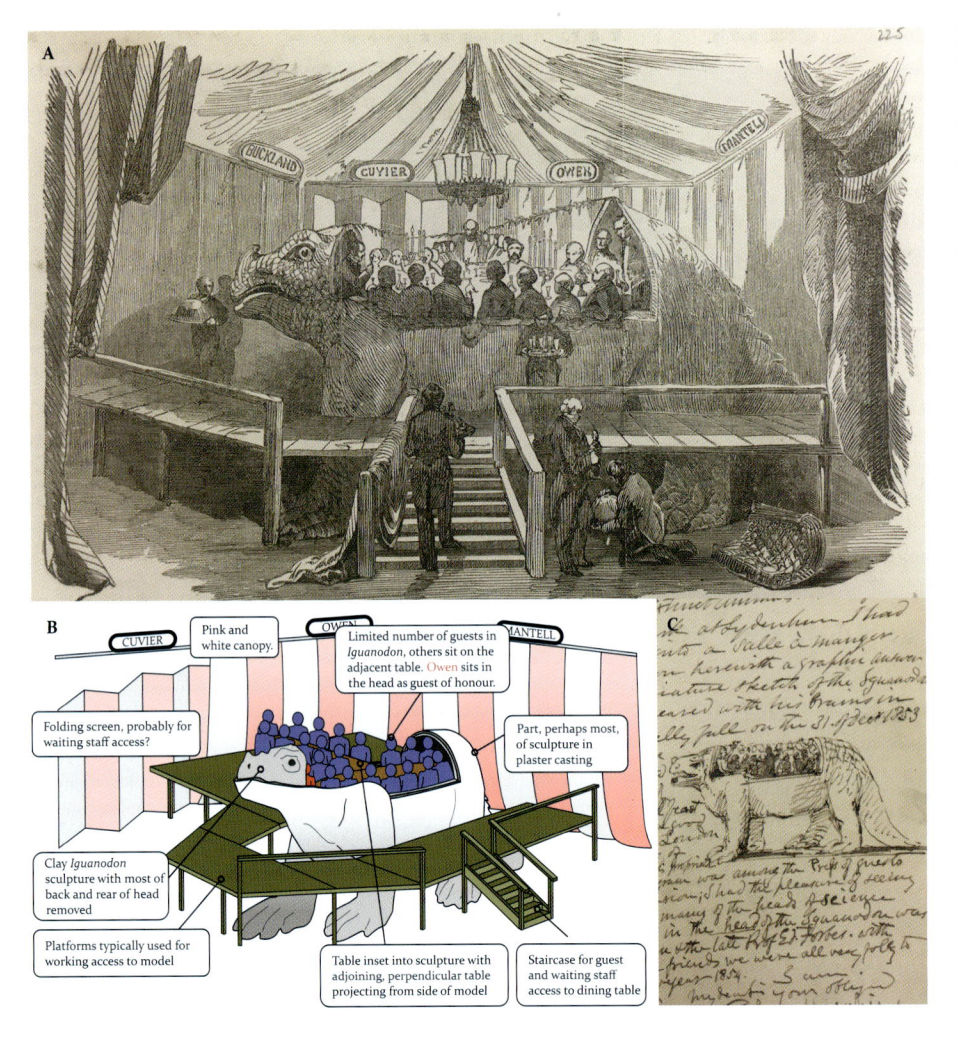

Fig. 3.24 The famous 1853 New Year's Eve *Iguanodon* banquet. **A)** 1854 lithograph from *The Illustrated London News*, the only drawing of the entire banquet drawn 'live' and likely capturing the most authentic version of events; **B)** cartoon interpretation of the dining arrangement based on contemporary drawings and accounts; **C)** 1862 drawing by Hawkins in a private letter – note the absence of the adjoining table.

geology and palaeontology – Owen, Conybeare, Forbes, Buckland, Cuvier and Mantell – were suspended (McCarthy and Gilbert, 1994).

Many details of the dinner are on record, including aspects of the seating plan, the poem or song about the event created by the naturalist Edward Forbes, and the content of the toasts. Of the latter, the theme was much self-congratulation, endorsement of Hawkins' reconstructions as authentic and accurate, as well as the celebration of Owen's genius and supposed supervision of the models.

But for all these details, some aspects of the *Iguanodon* banquet are uncertain. Exhaustive records of the event were not kept, and some sources of information contradict others. The number of diners, for example, is mostly stated as twenty-two (e.g. the 1854 write-up in the *Illustrated London News*; Hawkins' own description; McDermott, 1854), but was recorded as twenty-eight by Owen (1894). The question of guest numbers relates to the practicalities of fitting a dining party within the *Iguanodon* interior. The standing *Iguanodon* model is indeed vast, but could even the clay mould, with its more capacious interior, really house over twenty people *and* a dining table? Unbelievable as it seems, Hawkins' 1872 account indeed states that the entire dining party was crammed inside the 15-foot (4.6m)-long space in the back of the sculpture, with nine people along each side of the model and additional seats at both ends of the table for distinguished guests. Owen was given the loftiest position as the literal head of the table – a sketch by Hawkins produced on the night records his representation of the figurative brains behind the project (Bramwell and Peck, 2008). Owen's 1894 biography confirms this packed seating plan, but also mentions that the banquet layout included an adjoining table:

> The number of gentlemen present was twenty-eight, of whom twenty-one were accommodated in the interior of the creature, and seven at a side table on a platform raised to the same level.
>
> OWEN, 1894, P. 399

64 Chapter 3 – Bricks, iron and tiles: rebuilding the past

Though not widely reported, the presence of such a table is confirmed by an engraving published in the *Illustrated London News* (*see* Fig. 3.23; an image suggested by Cadbury (2000) as being a first-hand account of the event), as well as a report in the *Morning Chronicle* (McCarthy and Gilbert, 1994). These sources also suggest a less packed arrangement of eleven diners in the model, and ten at the table. Whatever the exact numbers, the impractical nature of housing so many people within the *Iguanodon* implies that the entire convocation only spent some of the evening inside it (perhaps only the toasts and other ceremonial moments?), and that they adopted a more comfortable seating plan for the rest of the evening (*see* Fig. 3.24B). Guests and waiting staff accessed the *Iguanodon* using stairs and raised platforms located around it – probably the same structures used by Hawkins and his team to create the sculptures and that can be seen in certain illustrations of the workshed.

As a publicity stunt, the *Iguanodon* banquet paid off handsomely. Newspapers across Europe reported the event (Cadbury, 2000; Secord, 2004) and the *Illustrated London News* filled half a page with a large-format engraving and write-up (perhaps not coincidentally, the editor of the *Illustrated London News* – Herbert Ingram – was in attendance at the banquet; Secord, 2004). It was surely not coincidental that the same page also featured a well-illustrated piece about Owen's remarkable work on reconstructing the moa. This was not 'news' in any real sense – Owen's *Dinornis* prediction was fifteen years old at this point – but it served the interest of the Crystal Palace Company, reminding readers of the restoration method at the heart of the Geological Court project and their consultant's intellectual prowess (Dawson, 2016). Along with a flurry of other publicity – more articles, posters and leaflets – such write-ups set the stage for the sort of spectacles and extravagance that the public could expect at the soon-to-open Geological Court and wider Park.

Hawkins, Ansted, Campbell and their teams would continue working on their palaeontological and geological creations until 1855. They achieved much of what was planned for the Primary and Secondary regions of the park but, while Hawkins at least finished all the planned Secondary Island palaeontological sculptures, no one area received the full intended complement of Geological Illustrations. After the park opened in June 1854, work continued on the Geological Court under the watchful eye of an enthusiastic public. Having finished his Secondary Island models in spring 1855, Hawkins and his team began work on what had

Fig. 3.25 One of the several versions of the invitations created by Hawkins for his famous New Year's Eve *Iguanodon* banquet. More complex versions, which included details of the workshed interior, were too difficult for Victorian printers to replicate.

once been their primary objective: the Tertiary Island sculptures, starting with a mammoth and moa.

So confident were the park authorities that these sculptures would be implemented that several Crystal Palace guidebooks and artworks included them (*see* Fig. 2.3; Doyle and Robinson, 1993; McCarthy and Gilbert, 1994), but underwhelming financial returns from the first year of opening saw all work on the Geological Court halted in September 1855. The cessation of construction was so sudden that even the half-completed mammoth and *Dinornis* were not finished and Hawkins was relieved of his post, never to resume work on the abandoned models. We'll return to this event in Chapter 12, after first familiarizing ourselves with the details of what Hawkins and his team of artists achieved.

Chapter 3 – Bricks, iron and tiles: rebuilding the past   **65**

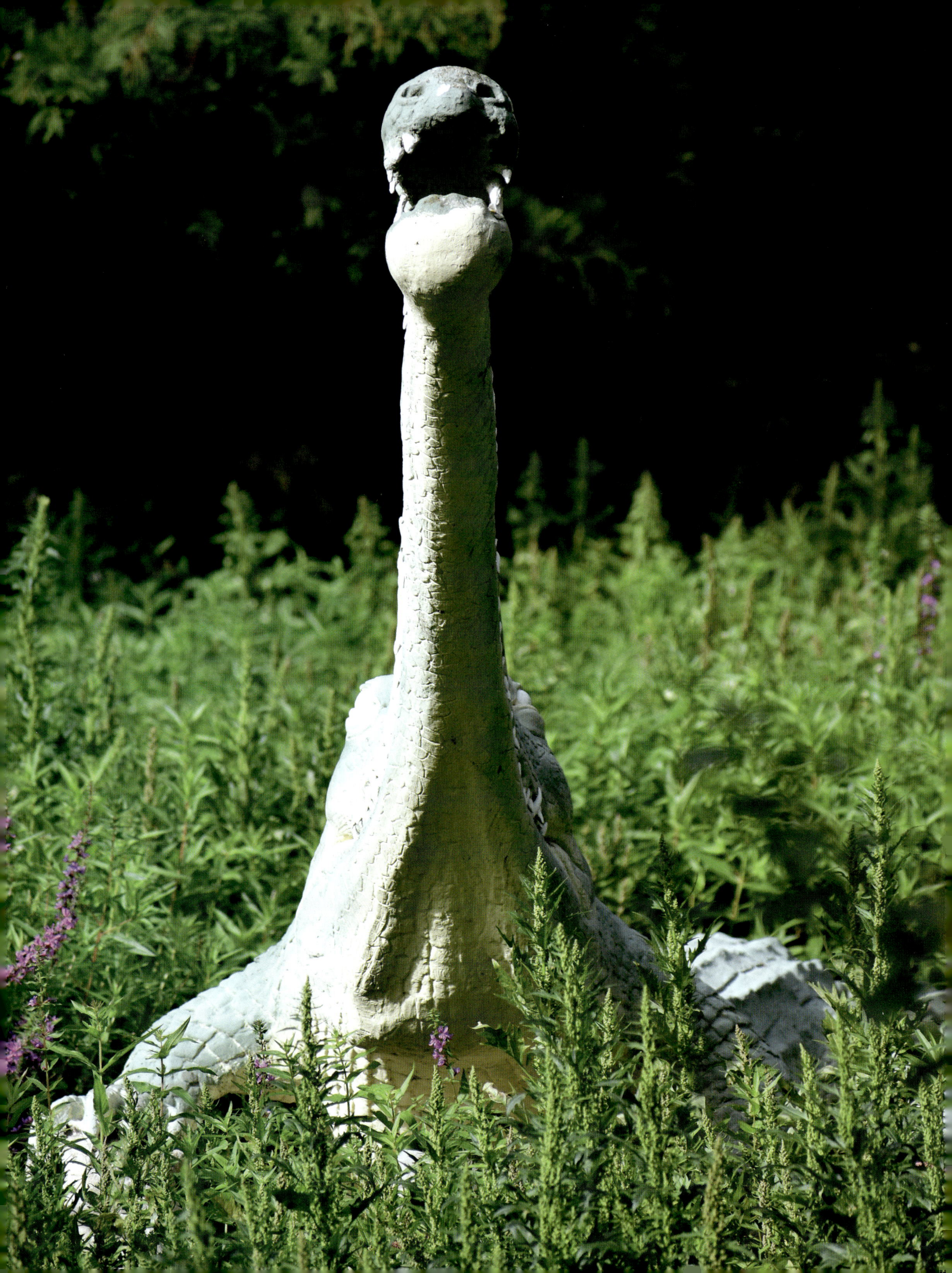

PART 2

Animals long since extinct

The footmarks, the bones, the very skin in some cases, of animals long since extinct, have been preserved by being buried in mud which has afterwards been converted into solid rock. From these obscure guides, the comparative anatomist has ventured to describe the general form, the habits, and the peculiarities of the race. From such descriptions, penned chiefly by Cuvier, Mantell, and Owen, has Mr. Waterhouse Hawkins restored and by degrees built up the animals. Possessing a great knowledge of the peculiarities of many living species; and being strong in his own feeling of what was probable and natural in the numerous details that required consideration, he has skilfully and cautiously constructed these restorations, and his embodiments of the opinions of the greatest palæontologists are indeed equally bold and conscientious. Professor Owen, the most eminent living authority upon these subjects, has kindly rendered Mr. Hawkins every assistance in his undertaking.

SAMUEL PHILLIPS, 1856, P. 196

The following chapters are dedicated to what are, for many, the main draw of the Geological Court: Waterhouse Hawkins' vintage reconstructions of extinct animals. We have followed the convention of certain guides (Owen's *Geology and Inhabitants of the Ancient World*, McDermott's *Routledge's Guide to the Crystal Palace and Park at Sydenham*) in covering the geologically youngest species of the display first, and then moving to progressively older animals. This is also the order we recommend visiting the sculptures in, entering the Geological Court via the impressive *Megaloceros* arrangement and ending with a grand panorama of the Secondary Island to take in the *'Labyrinthodon'*, *Dicynodon*, marine reptiles, dinosaurs and several Geological Illustrations. In these chapters, we have attempted to delve into the palaeoartistry of Hawkins' models, uncovering the sort of fossil material and modern reference species he and Owen had to work from in an effort to understand why their reconstructions were created as they were. We have also highlighted and explained where modern understandings of ancient biodiversity have altered the very names we have traditionally given the Crystal Palace dinosaurs. Many species restored by Hawkins are no longer considered valid (i.e. their taxonomic names are no longer considered 'correct' for various reasons and have fallen out of use; invalid names of this sort are indicated by quotation marks), others have acquired new names, and some are chimeric assemblages of several species. Modern life reconstructions of the species represented at Crystal Palace have been prepared to compare with those produced in 1854, but their modernity does not necessarily mean they are 100 per cent accurate takes on the life appearance of their subjects. Despite over a century and a half of additional science, there is still much we do not know about the palaeobiology of the animals represented at Crystal Palace, and these modern illustrations will also become dated as new discoveries and novel research approaches shed additional light on the realities of long-extinct species.

CHAPTER 4

The sculptures: mammals

Our exploration of Hawkins' Crystal Palace palaeoart begins with his mammals, a collection of sculptures mainly based on the Tertiary Island in the northeast region of the Court. These statues blend spectacular visuals and sterling demonstrations of Hawkins' anatomical mastery with a sorry history. Not only was the planned extent of mammal sculptures never achieved (see Chapter 12) but the executed models are the least documented, least discussed and most dilapidated of the palaeontological restorations. Several bear obvious hallmarks of neglect, low-quality repair work and some have gone missing altogether. The mistreatment of the Crystal Palace mammals is not new. They were omitted entirely from Owen's 1854 guidebook despite his extensive studies on fossil mammals, and visitors had to make do with the brief, sometimes incomplete, overviews of the mammal fauna provided in different iterations of the general park guides. Although well-featured in illustrations of Hawkins' workshed (see Figs 1.12, 3.20–21), the installed statues were rarely captured by artists looking to record the content of the Geological Court (see Fig. 3.17), and historic photographs are also relatively rare. We are forced to rely on any and all accounts of their existence for critical historic details, no matter how brief or imperfect, to ascertain facts as basic as how many sculptures were originally built and installed.

The seven mammal species restored by Hawkins were all 'classic' extinct taxa even in the 1850s. The Giant Deer *Megaloceros*, the giant ground sloth *Megatherium*, the three species of the tapir-like *Palaeotherium* and two species of the somewhat camel-like *Anoplotherium* were among the very first fossil vertebrates studied under 'modern' scientific mindsets in the late 1700s and early 1800s, and their interpretation informed founding components of palaeontological theory, including the reality of extinction. Their skeletal anatomy was well understood by the 1850s with most species being represented by complete, or near complete, remains, and excellent skeletal reconstructions had been created in museums and scientific papers, including several by Georges Cuvier. They were also no strangers to palaeoart, having been featured in the genre for several decades (see Figs 2.9, 2.14, 4.1). Owen was a noted authority on the anatomy of mammals and had worked on species featured at Crystal Palace, including *Megatherium* and *Megaloceros*.

These factors, combined with the fact that Hawkins was a recognized expert in depicting mammalian form (his post-Crystal Palace career involved writing a series of books on depicting mammal anatomy in art), put the Geological Court mammals in a very different reconstruction category to the more famous reptiles. With relatively little bony anatomy to estimate, and excellent guidance from living mammals for details of soft-tissue anatomy, guesswork was

Fig. 4.1 Mammalian palaeoart contemporary to the opening of the Crystal Palace. **A)** one of the first reconstructions of *Megaloceros giganteus*, shown in front of a mammoth and mastodon, from *Die Gartenlaube* (1854); **B)** *Palaeotherium* and *Anoplotherium commune* from *Peter Parley's Wonders of Earth, Sea and Sky* (1840).

largely restricted to elements of muscle bulk and skin texture. Hawkins' restorations have accordingly stood the test of time amazingly well, even being used to depict the life appearance of certain fossil mammals in recent texts (Prothero, 2016). Freed from the interpretative problems facing other subjects of the Geological Court, the Crystal Palace mammals show that Hawkins' palaeoartistic skills were the equal of later palaeoart masters when he had more than scraps of bone to work with. They demonstrate without question that the much remarked-upon 'wrongness' of certain Crystal Palace sculptures had more to do with the challenges of interpreting poorly known fossil species than the abilities of their artist.

'*Megaceros*', the Irish Elk

Visitors approaching the Geological Court from the northeast are greeted by one of the most impressive arrangements of the entire display: three regally-posed *Megaloceros giganteus* that, even though only slightly elevated from the surrounding pathways, tower over guests with their impressive 3m antler spreads, 2.8m-long bodies and 2m shoulder heights. They are easily the most spectacular mammal sculptures in the Court and the imposing stags are so stately that they would not look out of place amongst grand governmental buildings or atop plinths in city squares. Even the reposed doe, situated at the feet of the males amid the vegetation of the display, has the quiet dignity reserved for large, spectacular mammals.

Fig. 4.2 A deer of many scientific names, but introduced to Crystal Palace visitors as the Irish Elk: *Megaloceros giganteus*. If approached from the northwest, this spectacular arrangement of Giant Deer greets visitors to the Geological Court. Note the differential weathering on the doe, highlighting the distinction between metallic and concrete elements. (2017)

Representing recently extinct Pleistocene-Holocene (and, thus, essentially 'Ice Age') species, the Giant Deer serve as a perfect introduction to reconstructed extinct life, blending both the familiar features of living cervids with a definite otherworldliness: their giant size and unprecedented antlers. As noted above, they also leave no doubt about Hawkins' ability to create believable animal restorations, a fact that groundtruths his later, more exotic reconstructions. This said, it's perhaps also true that the grandeur and strangeness of dinosaurs, marine reptiles and giant sloths ultimately overshadow the deer somewhat, so their splendour is best

Fig. 4.3 The Crystal Palace *Megaloceros* on display in 1864 with their original fossil antlers – note the many differences with the strangely flat artificial antlers they bear today.

experienced before encountering the more exotic members of the Geological Court.

Megaloceros was a historic fossil species even to Hawkins and Owen thanks to the discovery of its remains dating to the late 1600s. By the early 1800s, enough material was known to reconstruct entire skeletons (e.g. Cuvier, 1827) and, with *Megaloceros* fossils relatively abundant in Pleistocene and Holocene localities, many British museums had excellent collections of this species. Given the abundance of information on *Megaloceros* and its long research history even in the 1850s, it is surprising that few efforts to restore the life appearance of Giant Deer seem to have been made before Crystal Palace (*see* Fig. 4.1A). Perhaps, in being found over a century before palaeoartistic reconstructions became fashionable, it was sidelined as a potential palaeoart subject by the exciting and exotic fossil reptiles and mammals being discovered in the early nineteenth century.

This wealth of *Megaloceros* fossils would have given Hawkins an excellent insight into its form and complemented his already excellent knowledge of deer anatomy, which was sufficient to inform a book dedicated to this topic (Hawkins, 1876). Combined, these points suggest that imagining *Megaloceros* anatomy was probably not a great challenge for Hawkins and his team. They created *Megaloceros* using an internal metal frame that supported bodies constructed of brick and concrete, while the heads and legs were cast in metal so as to capture the details of their faces, limb joints and feet. It's no great surprise that the *Megaloceros* are the most scientifically credible of all the Crystal Palace artworks, but whether they show Hawkins' palaeoart abilities at their maximum extent is another question. It was perhaps his ability to comprehend lesser-known and wholly unfamiliar fossil species that place him among the great masters of the genre, even though these insights are associated with sculptures that are more scientifically dated.

What we see of the Geological Court *Megaloceros* today is not their original configuration. Photographs from 1864 reveal the head of a second resting doe sculpture behind the low slope housing the two males, bringing the total of manufactured *Megaloceros* to four. Further details of this sculpture are scarce, but its disappearance may have something to do with a temporary nineteenth-century relocation of the *Megaloceros* to the Primary Strata. A private 1874 letter from Hawkins exclaims 'the gigantic Irish Elks have been placed on the Coal Measure!' and expresses disbelief at the disregard for the layout of the carefully planned Geological Court by the 'supposed guardians of the Crystal Palace property' (Brierley *et al.*, 2018). At least one *Megaloceros* never made it back from this trip, as its foot was recovered from the mine in 2001: this may represent the missing doe.

Conversely, it seems that a small, deer-like sculpture generally interpreted as a *Megaloceros* fawn (e.g. Doyle and Robinson, 1993; McCarthy and Gilbert, 1994) was not part of the original display. Historic newspaper accounts and park guides (e.g. McDermott, 1854; a January 1854 *Observer* report) discuss the composition of Hawkins' *Megaloceros* in relative detail but no fawn sculptures are mentioned. Several lines of evidence, outlined in more depth below, point to this sculpture representing an overlooked '*Anoplotherium gracile*' added to the *Megaloceros* display at an unknown date, under the mistaken assumption that it represented a juvenile deer.

Modifications have also taken place to the *Megaloceros* appearance. The stags' antlers were originally genuine fossils (*see* Fig. 4.3), which were available in relative abundance from Pleistocene and Holocene deposits across Europe and Asia, and especially so in bogs of Ireland and the Isle of Man. At some point in or before the 1950s these were replaced with lighter artificial structures (the current versions are fibreglass) that lack the upwards-projecting tines and impressive complexity of real *Megaloceros* antlers. This is just one of many examples of questionable conservation work that has befallen the Geological Court mammals. Paint analysis by Hirst Conservation (2000) has revealed no details of the original

Fig. 4.4 Likely reference fossils and inspirational animal species for *Megaloceros giganteus*. Osteological drawings from Cuvier (1827).

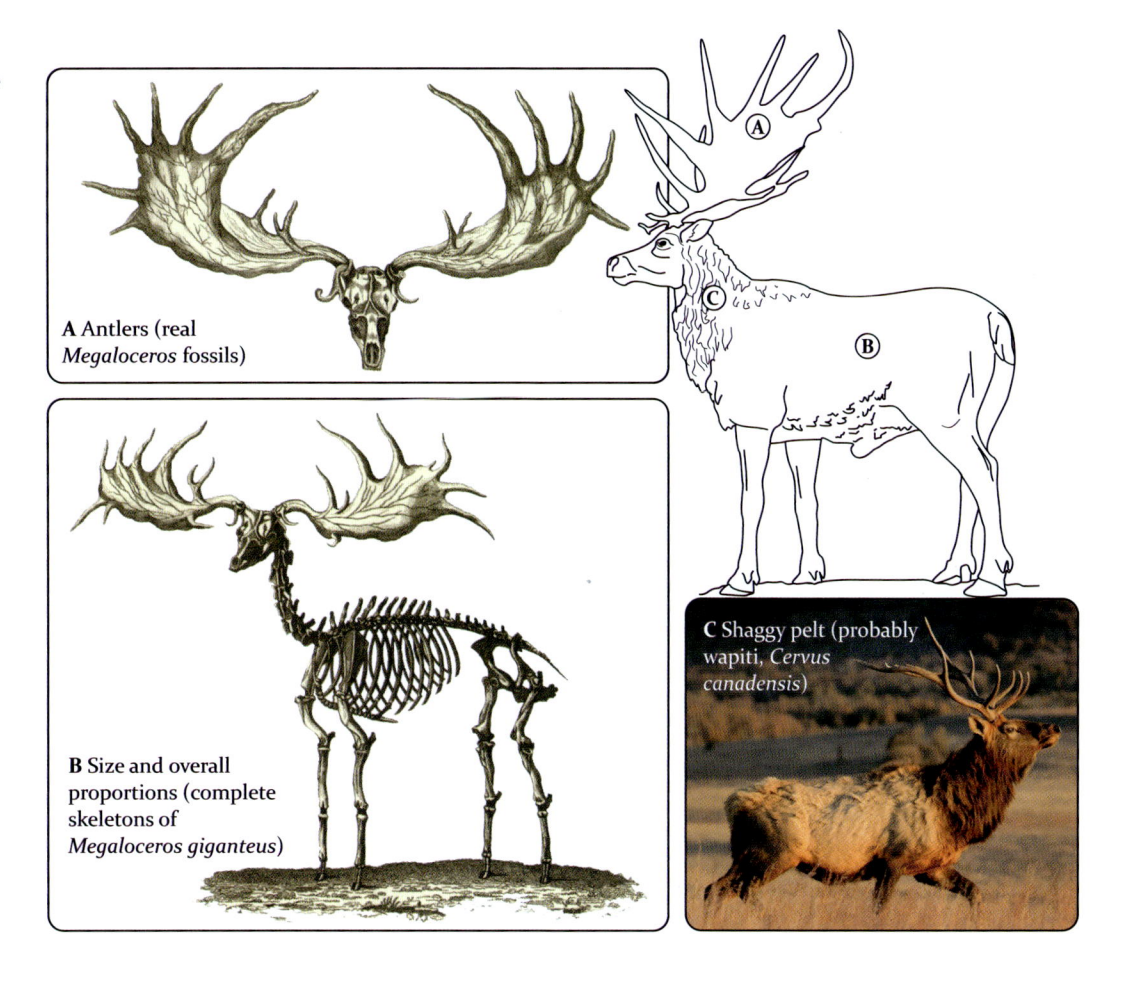

A Antlers (real *Megaloceros* fossils)

B Size and overall proportions (complete skeletons of *Megaloceros giganteus*)

C Shaggy pelt (probably wapiti, *Cervus canadensis*)

Megaloceros colouration, but red/brown or buff/straw colours seem to have dominated their subsequent history.

Megaloceros is unusual for an extinct animal for having several common names, including 'Irish Elk' (despite neither being strictly Irish nor related to any living deer we call an 'elk') and 'Giant Deer'. It has also had a variety of scientific names thanks to a complex taxonomic history, which ultimately ended up influencing some details of Hawkins' restoration. Initially interpreted as a Eurasian elk/moose-like animal and thus a 'New World' deer, both Cuvier and Owen realized that the Giant Deer had greater affinity to the cervine, 'Old World' branch of deer evolution. It was accordingly regarded as a member of the genus *Cervus*, with Owen coining the name *Cervus (Megaceros) hibernicus* in 1846. *Megaloceros* was actually the older, and thus correct name given rules of zoological nomenclature, but Owen's '*Megaceros*' became widely used for the next 100 years. Through this popularity, '*Megaceros*' almost established itself as the accepted generic name for this deer, until *Megaloceros* re-entered common usage in 1945. For a time, both names were equally applied. After decades of confusion, Giant Deer expert Adrian Lister (1987) finally made a call on which name should win out, establishing *Megaloceros* as the most appropriate on grounds of both nomenclatural priority and its revived usage.

This may seemingly be a matter of zoological minutiae – how can a name influence a palaeoartistic reconstruction? – but the idea that *Megaloceros* was a large-bodied member of *Cervus* may explain why Hawkins appears to have referenced several *Cervus* anatomies in his sculptures. These include thick neck manes, deep fur over the withers, the short, blunt tails and a line of long, shaggy fur along their bellies. These are especially obvious on the stags, but also present on the doe. Manes are not common to many deer females and we might assume that Hawkins was referencing the winter appearance of certain elk subspecies ('elk' as in the wapiti *Cervus canadensis*, not the Eurasian elk/moose). He may have thought that the shaggy appearance of a winter elk was apt for an Ice Age animal, or else that longer fur would look more obvious on his sculptures. In any case, his reconstruction stands out against the later tradition of reconstructing Giant Deer as a giant version of the red deer *Cervus elaphus* – a look that dominated twentieth-century *Megaloceros* palaeoart.

Hawkins' *Megaloceros* are fine representations of this animal but have been superseded thanks to discoveries giving us more detailed insights into their soft-tissue anatomy, life appearance and lifestyle. Although often regarded as a classic 'Ice Age' species, *Megaloceros* was actually a denizen of open temperate habitats and lived south of the steppes and glaciers we associate with the Pleistocene (Lister, 1994). It was also among the fastest of all deer and seems to have been adapted for long-distance, sustained running (Geist, 1999). *Megaloceros* was also probably more closely related to the fallow deer, *Dama*, than *Cervus* (Lister, 1994), which implies some differences in facial anatomy and colouration to *Cervus*-based reconstructions, as well as distinctions in particulars of fur and soft-tissue distribution. *Dama* have bulging throat regions and brush-like genital sheaths, both of which may have been present in *Megaloceros*.

At least fifteen, and maybe as many as forty depictions of *Megaloceros* are also known from Palaeolithic cave art (Geist, 1999; Guthrie, 2005) and these record further anatomical insights, including a shoulder hump thought to represent a fat store (Geist, 1999) and some details of colouration or patterning. Opinions differ on how to interpret *Megaloceros* colour from the monochrome line art of ancient humans, however, and we cannot regard the colours and patterns of this species as well understood despite these data. Some cave artworks show dark colours at the shoulder hump which taper into dark stripes across the body and neck, with a further dark stripe located behind the head. Another vertical stripe crosses the haunches, potentially signifying a white rump – a feature consistent with *Dama*. Geist

Fig. 4.5 Details of the Geological Court *Megaloceros giganteus* stags and doe in 2013, 2018 and 2021. Much of the fine detail in these sculptures reflects metalwork, rather than cast concrete.

Fig. 4.6 *Megaloceros giganteus* restored in 2021: a supreme runner adapted to live in open habitats. Hawkins' excellent anatomical knowledge of deer and our early grasp of the anatomy of *Megaloceros* fossils means modern reconstructions only differ in minor details of assumed soft tissues and colour from those at Crystal Palace.

72   Chapter 4 – The sculptures: mammals

(1999) interpreted these details as *Megaloceros* being pale all over with a dark stripe along the lower body and dorsal midline, while Guthrie (2005) portrayed more regionalized colouration with darker hindquarters grading into paler hues around the shoulder, framed by prominent stripes. Lister (1994) was more conservative, merely noting the likelihood of the dark shoulder hump. A further alternative is that the 'stripes' of Palaeolithic *Megaloceros* art are not stripes at all, but demarcation of colour regions.

However we interpret these ancient artworks, this information was entirely unknowable to Hawkins as the discovery of ancient European cave art post-dated the Crystal Palace project by over a decade, and they were not accepted as the authentic work of Palaeolithic humans until the early twentieth century.

Megatherium americanum

At the eastern tip of the Tertiary Island is the towering figure of *Megatherium americanum*, a famous giant ground sloth species from Pliocene-Pleistocene deposits of South America. The elephant-sized giant sloths are some of the most unusual and charismatic of all extinct mammals, and they are famous for their powerful, strongly clawed limbs and suspected capacity to prop themselves up on their legs and browse from the tops of trees.

Situated on a base of simulated Argentine glacial gravel, sitting on its hindquarters and rearing 4–5m into the air, Hawkins' *Megatherium* captures all these qualities. Although lacking the stately pose of the *Megaloceros*, it is undoubtedly an impressive sculpture. Its strangeness leaves visitors in no doubt that they have metaphorically stepped back in time, and its size and peculiar anatomy are more striking than that of the unusual, but not spectacularly fantastic *Palaeotherium* and *Anoplotherium*. Many details of the sculpture – especially the long, shaggy hair, foot anatomy (*see* below) and facial features – are exquisite, so it is disappointing that they are typically obscured by vegetation, sometimes leaving only the shoulders and head visible from public walkways. Decked out today in a brown-orange colour with dark brown eyes, the sculpture was battleship grey during the 1990s (Doyle and Robinson, 1995), but Hirst Conservation's (2000) paint analysis suggests it was originally a buff-yellow colour.

The *Megatherium* is unique in several respects. It was constructed through carving limestone blocks (Doyle, 2008) – specifically, Bath Stone – and was posed around a once living tree, the arms gripping the trunk and the head facing upwards as if reaching into the canopy. This is the only example in the Geological Court of a sculpture directly interacting with a real plant and, despite the apparent dominance of the sloth, it's actually the tree that has proven the more assertive of the two. Now long dead and pruned back, at one stage the growing tree broke free of the sloth's grasp by snapping off the left forearm and possibly both hands. The left arm and hand have been replaced with a fibreglass replica but, at time of writing, the right hand is missing. Whether this represents the loss of original sculpture material remains to be seen. The hands associated with the model in recent decades (including the existing left hand and broken hands known from photographs – *see* McCarthy and Gilbert, 1994) are undersized, lack claws and are crudely carved. Although some sources (e.g. McCarthy and Gilbert, 1994) regard these as representing Hawkins' original work, their poorly realized anatomy and basic construction suggest they are more likely replacements of now-lost original components.

Fig. 4.7 Rearing some 4–5m high is Hawkins' *Megatherium americanum*, shown here as it was in 2013. Vegetation frequently obscures this sculpture and the expertly sculpted feet, legs and tail have been difficult to observe in recent years.

Fig. 4.8 Details of the Crystal Palace *Megatherium americanum* from 2013 onward. Note the finely sculpted hair, the excellent rendering of the crouching hindlimbs and inturned feet, and also the strangely small left hand. The right hand is missing.

There is no doubt that Hawkins knew *Megatherium* should have large clawed hands and he almost certainly outfitted his model with them. They appear in an 1853 image of the model during construction (*see* Fig. 3.20B) and were remarked upon in a January 1854 *Observer* article on the Crystal Palace's *Geological Wonders*: 'his forefeet are about a yard in length, and terminated by powerful claws'. Hawkins' illustrations of *Megatherium* from the 1850s and '60s (including artwork associated with Crystal Palace) also show it with correctly restored hand anatomy (*see* Fig. 12.6). It seems likely, therefore, that the tiny, blunt hands of recent history are inaccurate replacements of Hawkins' originals. Further conservation work of a nature more sensitive to the original form of the model was recently carried out to repair the sloth's broken tail, which had to be completely recreated after the hair-like texture added by Hawkins was worn off. The foundations of the model have also had to be repaired and partly replaced.

The Crystal Palace *Megatherium* is among the oldest life restorations of a ground sloth. Though several skeletal reconstructions were published prior to the 1850s and some simple life restorations were performed (e.g. Fig. 2.9), few detailed efforts had been made at reconstructing these charismatic animals before Crystal Palace. This is despite the research history of *Megatherium* beginning nearly six decades before Hawkins commenced work on his model. Georges Cuvier had described its anatomy at length during the late 1700s and *Megatherium* had the honour of being the first fossil skeleton mounted in a museum – specifically, the Museo Nacional de Ciencias Naturales, Madrid – in 1795 (Rudwick, 2005; Argot, 2008). These accolades saw *Megatherium* become a common fixture in palaeontological texts of the early 1800s,

Fig. 4.9 Likely reference fossils and inspirational extant species for the *Megatherium americanum* sculpture. *Megatherium* skeleton from the nineteenth-century zoology print series *Iconographia Zoologica*; *Glossotherium* from Owen (1842b).

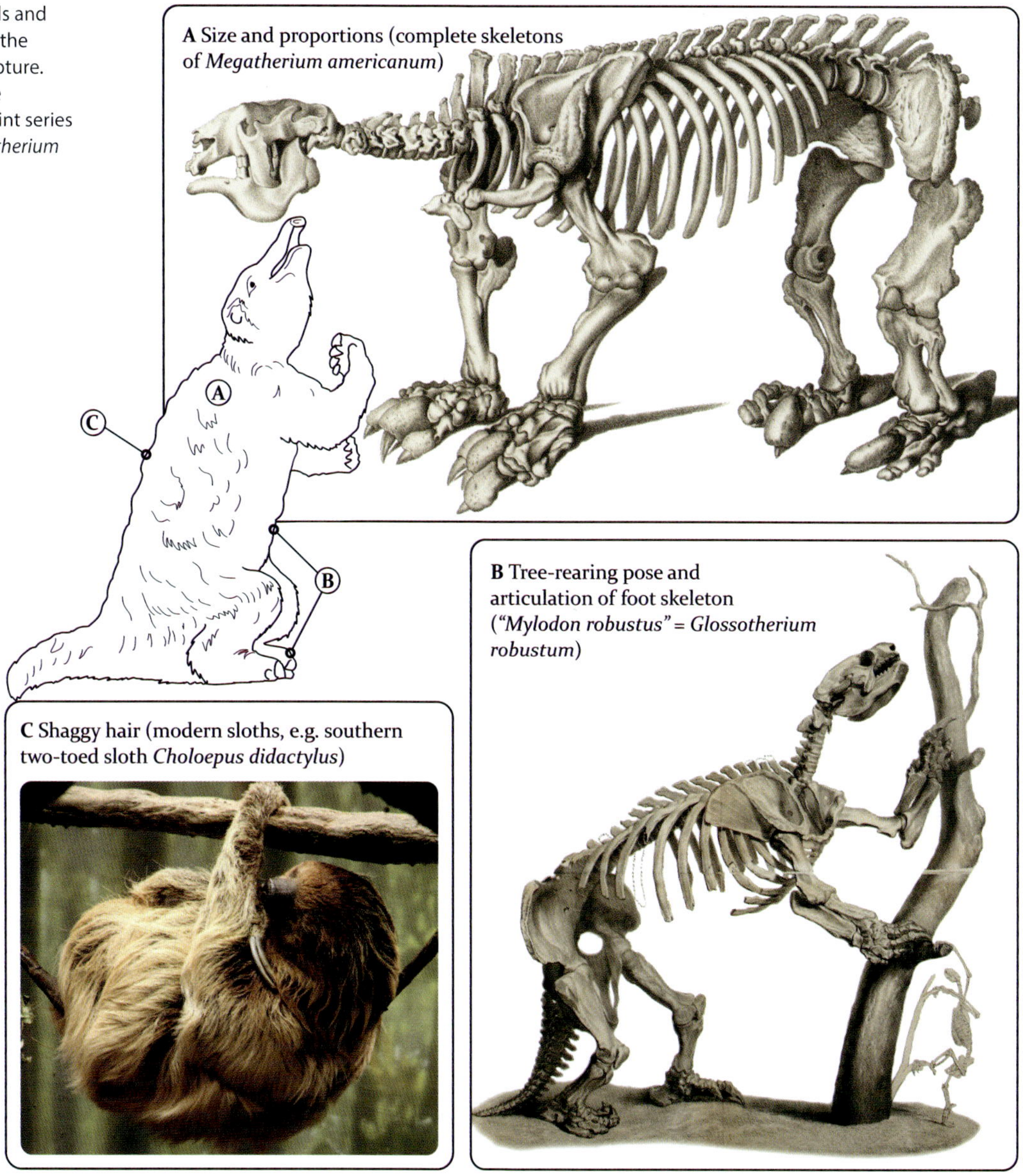

A Size and proportions (complete skeletons of *Megatherium americanum*)

B Tree-rearing pose and articulation of foot skeleton ("*Mylodon robustus*" = *Glossotherium robustum*)

C Shaggy hair (modern sloths, e.g. southern two-toed sloth *Choloepus didactylus*)

with most treatments repeating Cuvier's interpretation of it as a flat-footed, trunked quadruped with particularly dextrous forelimbs (Rudwick, 2005; Argot, 2008).

Scholars were less certain about its habits, however, and even into the 1850s the function of its powerful limbs and claws were debated: were they used for burrowing, climbing or manipulating vegetation (Argot, 2008)? Owen's studies on another ground sloth, '*Mylodon robustus*', shone much-needed light into aspects of ground sloth lifestyle and anatomy (Owen, 1842b). This species, known today as *Glossotherium robustum*, stood on the sides of its feet so as not to blunt its large toe claws, and had a complex means of walking on its hands, bearing its weight on its shorter, clawless fingers and the knuckles of longer, clawed digits. He viewed giant sloths as browsers and tree fellers based on numerous lines of anatomical evidence and, under his direction, this bear-sized sloth was restored as a tree-rearing biped at the Hunterian Museum (*see* Fig. 4.9B). Owen would eventually publish similar anatomical and lifestyle interpretations for *Megatherium*, but not until 1861 – several years after the opening of the Crystal Palace.

It's in this context that Hawkins constructed his *Megatherium* rearing into a tree, balanced on its tail and the sides of its feet. Whether he simply projected Owen's *Glossotherium* work to this larger sloth or Owen gave him advance notice of ideas he would eventually publish on *Megatherium* is unknown, but Hawkins' rearing, tree-grasping restoration was a radical rejection of the Cuvierian, quadrupedal

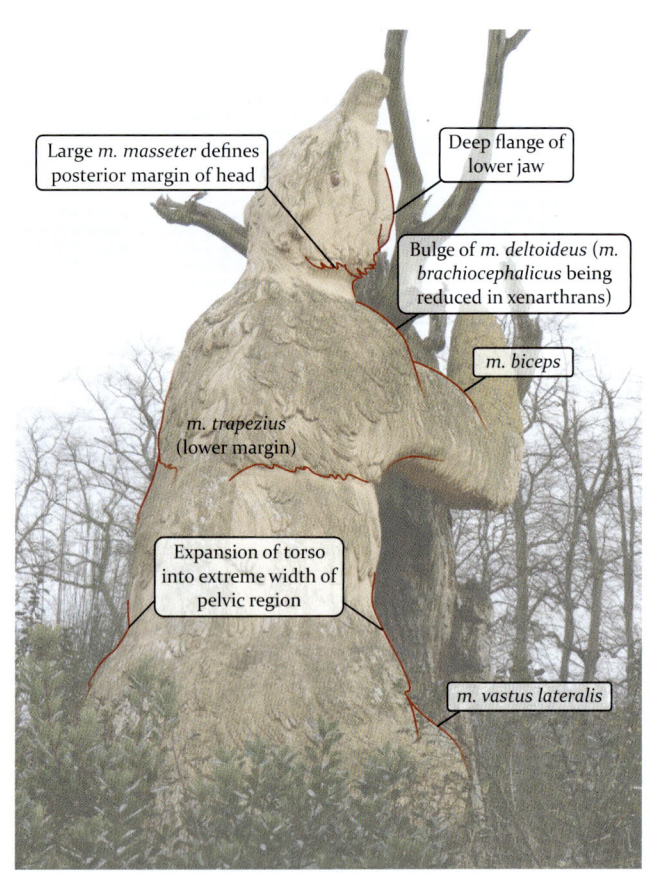

Fig. 4.10 Hawkins' *Megatherium* is a fantastic example of how, even through sculpted shaggy fur, he accurately captured muscle layouts and important anatomical details.

Megatherium of tradition. The depiction of such a powerful pose in a giant fossil species would have been novel to the public and scholars alike. It was not a stretch to imagine the bear-sized *Mylodon* of the Hunterian Museum rearing onto its back legs to browse tall trees, but showing an elephant-sized sloth doing the same was a bold statement about the anatomical sophistication and physiology of extinct animals (Argot, 2008). Moreover, as is demonstrated in many Crystal Palace models, Victorian scholars often imagined fossil animals as variants on recognizable modern forms – paleotheres as tapirs, teleosaurs as gharials, pterosaurs as birds and so on – but Hawkins' *Megatherium* was boldly different from anything alive today (Argot, 2008). This not only left visitors to the Geological Court in no doubt that prehistoric life could be of a thoroughly unfamiliar and exotic nature, but demonstrated the power of science to restore and interpret unusual fossil animals.

Despite its vintage, Hawkins' *Megatherium* has held up well as a portrayal of ground sloth form. It is not only proportionally accurate to modern ideas of *Megatherium* anatomy, but has a respectable depiction of musculature buried under its long, shaggy hair. Muscle bulges are picked out by the distribution of hair texturing and are especially visible around the shoulders, neck and head. The contrast between the powerful but relatively slender shoulders and the wide and robust pelvic region is also admirably captured. The feet are appropriately inturned and equipped with enormous claws on the leading toes, and the tail is stout and muscular. The tree-grasping pose – perhaps capturing the act of accessing canopy leaves, or shoving the tree over – accords entirely with modern interpretations of ground sloth behaviour. Such depictions have become stereotyped for *Megatherium* and other giant sloths in palaeoart, despite ground sloths probably spending much of their time on all fours, especially when travelling long distances (Melchor *et al.*, 2015).

The shaggy hair covering Hawkins' *Megatherium* was not based on fossil data, but borrowed from the long fur of living sloth species (*see* Fig. 4.9; Anonymous, 1854). This depiction anticipated the discovery of fossilized ground sloth hair – that of *Mylodon* – by almost half a century (Woodward and Moreno, 1899) and established the convention of portraying even gigantic fossil sloths with a thick fur coat. This is one aspect of Hawkins' *Megatherium* that might prove erroneous in future, as the restoration of four-tonne sloths with such extensive fur has been questioned on grounds of thermal energetics (Fariña, 2002). Models of giant sloth physiology imply that, like many large mammals alive today, staying cool was more of a challenge than staying warm, and that a coat of thick hair would have put them at near perpetual risk of hyperthermia (Fariña, 2002; Fariña *et al.*, 2013). It's possible that hairy coats were restricted to smaller ground sloths living in colder climates (the one- or two-tonne, high altitude *Mylodon* being a good candidate) and that larger species living in more temperate settings – including *Megatherium* – had reduced quantities of insulating hair. Some – though by no means all – artists have thus started to buck the centuries-old trend of shaggy *Megatherium* art started by Hawkins to restore this and other giant sloths with reduced or absent hair (*see* Fig. 4.11).

A final difference between Hawkins' *Megatherium* and our modern reconstructions is his inclusion of a short proboscis. This would not have shocked scholars of the time as a short trunk was predicted for *Megatherium* by Cuvier and also endorsed by Owen (1842b) based on the retracted rostral bones and expanded nasal cavity of its skull. Such features are among those characteristics associated with trunks or proboscides in living mammals (e.g. Wall, 1980) but trunked

Fig. 4.11 *Megatherium americanum* as we imagine it today, still rearing into trees. A key difference between some modern and Victorian interpretations of this species – other than the absence of a trunk – is the possibility of vastly reduced hair coverings owing to issues of body temperature regulation in very large mammals.

sloths were not destined to become a popular convention and quickly fell out of favour among palaeoartists. Recent studies have shown that Cuvier, Owen and Hawkins were not entirely incorrect in anticipating an unusual arrangement of facial tissues for *Megatherium*, however, even if evidence for a trunk is still lacking. Instead, megatheriines likely had extensive nasal cavities and large, prehensile lip tissues (Bargo *et al.*, 2006). While we thus cannot regard Hawkins' *Megatherium* trunk as correct, it is perhaps closer to reality than many subsequent reconstructions which have depicted this species with a minimized, tree sloth-like lip and nasal anatomy.

Palaeotheres

A set of mammal reconstructions with an especially tragic history are the palaeotheres, the relatively small, tapir-like sculptures located on the southwest bank of the Tertiary Island. Thanks to continued vandalism, poor conservation work and the loss of one model, they are some of the most mistreated sculptures of the Geological Court. Today, two small palaeotheres remain in the park but they were once joined by a much bigger, hulking individual that represented the largest known species of the *Palaeotherium* genus.

Fig. 4.12 The sitting sculpture of '*Palaeotherium minus*' (= *Plagiolophus minor*) alongside the standing *Palaeotherium medium*, two of the original three palaeothere sculptures made for the Crystal Palace Park. (2009)

Historic photos and film footage show that this trio was once visible from afar – even from the region of the Secondary Island – but they are now difficult to see owing to the close proximity of large trees and bushes.

The two surviving palaeothere sculptures represent different species. The slightly larger, standing one is *Palaeotherium medium*, while the smaller, seated model is '*Palaeotherium minus*' – today known as *Plagiolophus minor*. Although no-longer congeneric, both models still represent types of

Fig. 4.13 Records of the Crystal Palace *Palaeotherium magnum*, a model now missing from the park through unknown circumstances. **A–B)** *Pa. magnum* in the background of workshop illustrations, being stored behind and above the *Iguanodon* (arrowed in B); **C)** photographed in 1855 during the construction of the Geological Court; **D)** featured in a Geological Court promotional drawing alongside *'Plesiosaurus' macrocephalus*, from the 1853 periodical *Die Gartenlaube*; **E)** photograph of the model alongside *Plagiolophus*, from 1958 – also note the distinctive, now lost head anatomy of *Plagiolophus*.

palaeotherid: early members of the horse lineage. Despite somewhat resembling small horses, they were not closely related to modern equids. Their sculptures were realized using the same combination of metalwork and building materials anchored to concrete bases seen in *Megaloceros* and, like the Giant Deer, their bodies have seen substantial restoration in recent decades. At one point they were even removed from display owing to the considerable damage they had experienced. Prior to this, they were also moved along with the *Anoplotherium* to allow for development of a small petting zoo on the Tertiary Island in 1953.

While recent restoration works have attempted to faithfully repair the surviving palaeotheres, photographs of *Plagiolophus* from the mid-twentieth century show that it once had a very differently shaped head, now lost (*see* Fig. 4.13E). The current head, clearly derived from *Pa medium*, must have been installed based on *Pa. medium* in the latter half of the twentieth century. This modification not only removed a unique anatomical interpretation from the court but has been installed without close reference to the anatomy sculpted by Hawkins and his team, such that the head-neck join is now anatomically jarring. This

78   Chapter 4 – The sculptures: mammals

has been difficult to appreciate since 2014 when vandalism decapitated this model once again: thankfully, this second '*Pa. minus*' head is safely in storage so that it, if not a replica of the original, can be reattached during future conservation works. Currently painted a pale red-brown, which paint analysis indicates was likely their original colour, the palaeotheres were once coloured in one of the most striking schemes of the Geological Court: off-white with pink-orange horizontal stripes.

The presence and detailed appearance of the third, large *Palaeotherium* is well-documented in illustrations and photographs (*see* Figs 3.15, 4.13), although its absence from the park has been largely overlooked in previous overviews of the site. Evidently a large model – perhaps the size of a small horse – it represented the relatively large Palaeogene species *Palaeotherium magnum*, and was present in the park until at least 1958. Photographs show that it was once situated close to *Plagiolophus* in a standing pose, dwarfing its neighbour with a more muscular and robust physique.

McCarthy and Gilbert (1994) assumed this model represented a prior incarnation of the *Pa. medium* model, which was then modified in the late twentieth century, but this is unlikely to be true. Maps of the Geological Court created by Hawkins (*see* Fig. 12.7) as well as some editions of the Crystal Palace Park Guide (Anonymous, 1871) indicate that three palaeothere species once existed in the Geological Court and, moreover, obvious anatomical distinctions between the two sculptures – all aspects of size, proportion and robustness – leave little doubt that they were separate models. But while we can be certain that a *Pa. magnum* model once existed, its fate after 1958 is mysterious. Perhaps it was relocated or destroyed when certain mammal sculptures were moved for the petting zoo, or perhaps it met an unfortunate end at the hands of vandals? No remains of the model seem to be present in archives but the long, complex and variably documented history of the park gives hope that some remnant, or at least clues to its disappearance, await future discovery.

Today, palaeotheres are a type of fossil mammal largely known only to specialists, but in the nineteenth century they were standard inclusions in texts and educational materials related to earth sciences (*see* Figs 2.9, 4.1B). Recovered from the Eocene-Oligocene deposits of the Paris Basin at the turn of the nineteenth century, *Palaeotherium* was one of the first fossil mammals found in relatively old strata and was studied in detail by Georges Cuvier during the early 1800s (e.g. Cuvier, 1822). Thanks to the recovery of near-entire skeletons for several *Palaeotherium* species, their osteology and lifestyle as browsing herbivores was well understood from early on. Cuvier made specific predictions about their life appearance, which he thought would strongly resemble that of tapirs, despite their closer affinity to horses (Rudwick, 1997). Cuvier furnished his monographs with life reconstructions of these animals alongside drawings of their bones, making *Palaeotherium* the subject species for some of the oldest palaeoartworks (*see* Fig. 2.9; Rudwick, 1992, 1997). Cuvier's descriptions, specimen illustrations and life reconstructions were widely copied in European literature and would have furnished Hawkins with sound anatomical blueprints to base his sculptures on. Like *Megaloceros*, *Palaeotherium* was not a genus that would have especially stretched his skills at predictive anatomy.

Although representing three species from the same group, the Crystal Palace palaeotheres only share a basic body plan: in detail, they differ in a number of ways. Their shared features are rooted in the Cuvierian idea that palaeotheres would strongly resemble tapirs and each is thus equipped with short limbs, robust bodies, small, rounded ears and short trunks. They are not slavish reproductions of these living animals, however, as they have shorter, taller faces, higher set eyes, more gracile legs, longer tails and three toes on each foot (tapirs have four toes on their forefeet, and three on the hindlimbs). They also appear to be largely or entirely hairless, lacking the short fur and manes that characterize modern tapirs. Details of the faces and bodies originally differed between each sculpture to a greater degree than might be implied by their osteology; this might reflect Hawkins drawing reference from different living species to the extent of overriding details inferable from fossils.

Palaeotherium medium

The most tapir-like palaeothere at Crystal Palace is *Palaeotherium medium* (*see* Fig. 4.12). Reconstructed with metal legs and an aggregate body, many of the details seen on the modern sculpture are recent replacements: the ears, trunk, tail were all broken or missing until modern restoration work, and the body was weathered to the point of obscuring fine features. The 1.7m-long sculpture is largely intact today, although its exterior has been so badly denuded by encrusting plants and lichen that the restored colouration can barely be seen.

Palaeotherium medium is posed relatively neutrally, as if standing or walking slowly alongside the other palaeothere species. The sculpture gives the impression of being a relatively slow animal of forests and woodlands – largely how we still interpret *Palaeotherium* today. As implied by its name, this species was mid-sized for the genus and occupied this same distinction among the three palaeotheres of the Geological Court. Perhaps because this species was not as well represented by fossil material as the other palaeotheres reconstructed by Hawkins, it seems living tapirs – perhaps especially the Malayan tapir *Acrocodia indica* – were strongly channelled into its form, as is evidenced by the overall proportions, arched back, the posture of the neck and head, and the muscular, podgy torso with vertical flank creases. Further creases can be seen lining the back of the lower jaw, which imply that Hawkins imagined *Pa. medium* as having tough, stiffened skin. The face of the sculpture is relatively gracile and terminates in a slender trunk, as was considered appropriate for this species at the time.

Although the body of Hawkins' *Pa. medium* is still a relatively sound artistic interpretation, details of the head contrast with our current understanding of palaeothere anatomy. *Palaeotherium medium* had a somewhat longer, lower skull than is implied in Hawkins' reconstruction (Rémy, 1992) and it is doubtful that this species bore a well-developed proboscis. Although *Pa. medium* has generally been considered a trunked species by researchers and some authors have continued to use Hawkins' reconstruction to illustrate the life appearance of *Pa. medium* (e.g. Prothero, 2017), investigations into the likelihood of trunks in other fossil mammals suggest *Pa. medium* only bears a few features associated with the development of a proboscis (e.g. a retracted nasal cavity and strengthened upper jaw bones) and lacks many others (Wall, 1980). The presence of a few trunk-supporting features may imply a well-developed and mobile set of lips, but also suggests a more conservative facial condition overall than that present at Crystal Palace.

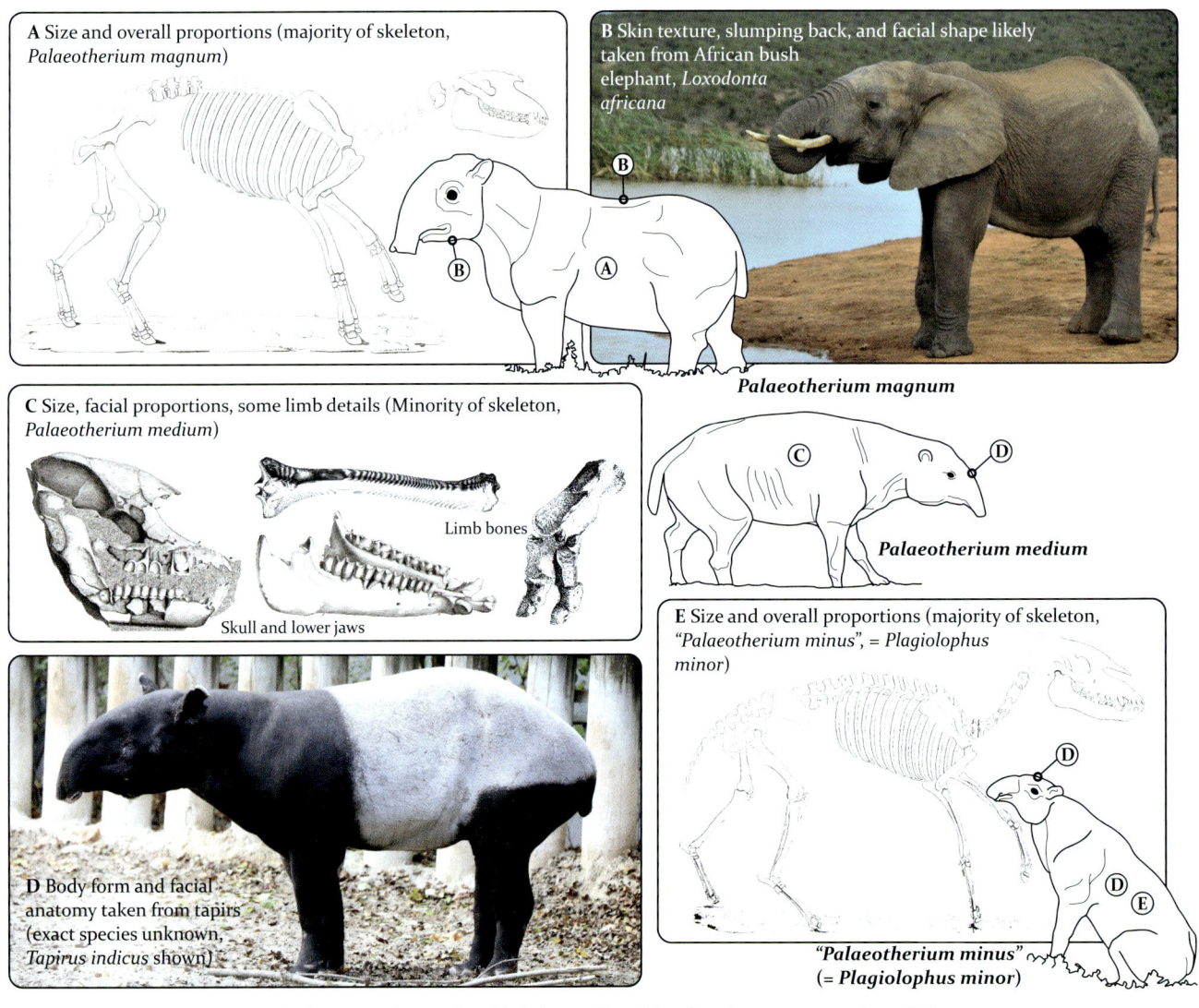

Fig. 4.14 Fossil and extant animal information that Hawkins likely factored into his palaeothere reconstructions. Skeletal illustrations from Cuvier (1822).

80   Chapter 4 – The sculptures: mammals

Fig. 4.15 Additional details of the Crystal Palace *Palaeotherium medium*. (2013, 2021)

'Palaeotherium minus'

Alongside *Pa. medium* sits Hawkins' take on *'Palaeotherium minus'*, a palaeothere now called *Plagiolophus minor*. The sculpture is of a sheep-sized animal that, like its standing counterpart, has suffered much damage and repair in recent years. Historic photographs show that this sculpture was once a short-faced creature of perhaps 1.3m total length, with large eyes and a stout, robust proboscis (*see* Fig. 4.13). At some point between 1958 and the 1990s the head was replaced with one cast from *Pa. medium*, and the different sizes, poses and anatomy of these sculptures seems to have created some difficulty in grafting the *Pa. medium* head onto the *Plagiolophus* neck. The posterior region of the cranium had to be removed and the neck lengthened to accommodate the repair, and the animal is now looking forward when it originally looked upwards. Continued vandalism has led to the sculpture losing its head twice more since this restoration job, most recently in 2014 when the decapitated part had to be fished out of the Tidal Lake. Perhaps because the Geological Court mammals are less conspicuous and popular than their reptile counterparts, damage of this nature often goes unrepaired for years.

Hawkins and his team had the majority of a skeleton to work with when restoring *Plagiolophus* (*see* Fig. 4.14) and reproduced its size and proportions faithfully. The body is much like that of *Pa. medium* although it lacks the flank creases and other skin details. Other than size, the sculptures chiefly differed in details of their faces. Unfortunately, records of the original *Plagiolophus* face are scant, but it evidently had a boxy, robust quality that contrasted with the somewhat longer, more gracile face of *Pa. medium*. The ears were relatively pointed, the eye large and the short trunk had a strongly convex, bulbous quality. The lower lip also seems to have been long, perhaps extending along much of the proboscis length. Overall, it seems to have been intermediate in anatomy between the restored faces of *Pa. medium* and *Pa. magnum*. The neck shows several strong muscle contours and, in its original guise, seems to have resembled the humped, muscular neck of the *Pa. magnum* sculpture more than that of *Pa. medium*. The tail is short, relatively slender and curves behind the body in a pleasingly naturalistic fashion: a great example of the subtle ways Hawkins introduced character to his sculptures.

As with *Pa. medium*, Hawkins' take on the *Plagiolophus* body remains relatively accurate to modern interpretations, but the presence of the proboscis and the shape of the face have fared less well over time. Hawkins was likely working from incomplete skulls and thus assumed that features seen in other *Palaeotherium* species were also present in *Plagiolophus*. Superior, near-complete skulls discovered since have shown that *Plagiolophus minor* had a relatively low, broad face (Rémy, 2004) with a narrow snout, and that it also lacked many of the features we correlate with trunks in living mammals (Wall, 1980). It thus probably had lips and nasal tissues more akin to those of a horse rather than a tapir, and this change would have dramatically altered the face compared to Hawkins' reconstruction (*see* Fig. 4.16).

Palaeotherium magnum

The loss of the *Pa. magnum* sculpture has passed virtually unnoticed in the history of the Crystal Palace park and the only details known about it are those which can be gleaned from historic photographs and illustrations (*see* Fig. 4.13). These at least give a general idea of its size and form. Our best record, a photo of the model from 1958, demonstrates a size appropriate for its species: about twice the dimensions of the sheep-sized *Plagiolophus*, and thus about 2m long from snout to tail. It was also the most anatomically divergent of

Chapter 4 – The sculptures: mammals **81**

Fig. 4.16 The life appearance of *'Palaeotherium minus'* – today known as *Plagiolophus minor* – as imagined in the twenty-first century. The skull of this species has few features indicating a developed trunk, so has been restored with more horse-like snout tissues.

the three palaeotheres, contrasting with the tapir-like forms of the other sculptures despite Cuvier's strong advocacy for such a life appearance in *Pa. magnum*:

> There is nothing easier than to represent this animal in its living state, for it is only necessary to imagine a tapir as large as a horse. A naturalist who would have taken the trouble to count its digits would indeed have found one fewer on the forefoot; if he had examined its molars, which so many naturalists today fail to do, he would have found still other differences. But for most people there would only have been that of size; and if one can count on the analogy, its hair was short, or indeed scarcely more than that of the tapir and the elephant.
>
> CUVIER, 1812 (TRANSLATED IN RUDWICK, 1997, P. 66)

Pa. magnum was not only the largest of Hawkins' palaeothere models but also the most robust, with thick legs, a proportionally large, deep head and a muscular body. It also had a concave back and a wrinkled skin texture. The proboscis was the most developed of the three palaeotheres and was depicted as being anchored far up the skull, creating a voluminous trunk that hung steeply over the lower jaw, terminating well below the lower lip. The eye was large and deeply set between a large brow and a creased, hollow-looking cheek region. The overall gestalt is not that of a horse-tapir, but of horse crossed with another trunked animal: an elephant, especially African bush elephant *Loxodonta africana* (see Fig. 4.14).

The introduction of elephant features into an ancient horse relative seems unusual and would be a highly questionable reconstruction decision if executed today. Elephants are, after all, very distant relatives of hoofed mammals and there is no reason to think that they are a good soft-tissue analogy for *Palaeotherium*. In the 1850s, however, linking *Palaeotherium* with elephants may not have seemed such a stretch. The broader taxonomic understanding of Mammalia in the mid-1800s was that horses and their fossil relatives belonged to the 'Pachydermata', an (artificial) group we today associate with 'thick-skinned' mammals such as elephants, rhinoceros and hippos. In the early 1800s, however, the group also included many hoofed mammals. Although it was still unconventional for Hawkins to incorporate elephant features into his *Pa. magnum* restoration, he may have felt he was still within a 'safe' degree of anatomical inference given the broad classification of 'pachyderms' among contemporary scientists. Furthermore, while Hawkins had a reasonable amount of skeletal material to work from with *Pa. magnum* (see Fig. 4.14), details where he has especially applied elephantine features – such as the torso and skull – were incompletely known to him, leaving him free to apply his own interpretations.

Hawkins' apparent adoption of elephant anatomy for his *Pa. magnum* made for a distinguished sculpture but

Fig. 4.17 A modern take on *Palaeotherium magnum*: a robust, mid-sized mammal with features of horses and tapirs. The extent of the soft tissues around the mouth in this reconstruction is tentative, as the skull of *Pa. magnum* bears some features of augmented lip and nasal anatomy, but not enough to infer a full proboscis. This issue has not been looked into in detail, however, and may warrant revision with further study.

has dated his reconstruction against our understanding of equoid life appearance more than the other Geological Court palaeotheres. Although a large animal for its group, *Pa. magnum* was not as robust as the Crystal Palace sculpture implied. In reality, it had gracile legs and a low, long skull. Thought to be a large browser of closed, vegetated environments (Rémy, 1992; Hooker, 2007), complete *Pa. magnum* skeletons suggest a form more like that of a stocky horse than a diminutive elephant (*see* Fig. 4.17) and also show that the concave back and proportionally large head were over-speculative deviations from palaeothere form. Several details of the head have also not been borne out by discoveries of more complete *Pa. magnum* skulls (Rémy, 1992). In addition to being long and low, not high and short, they also lack many adaptations associated with trunk development (Wall, 1980). The especially large, low-hanging proboscis of the Geological Court *Pa. magnum* is thus likely erroneous, although a set of fleshy lips may be apt for this species.

'Anoplotheres'

The traditional view of the Geological Court is that three *Anoplotherium commune* statues represent the last mammals of the Tertiary Island, their location at the western shore placing them closest to the reptiles of the Secondary region. Some confusion exists, however, over the specific identification of these sculptures, the resolution of which complicates not only our interpretation of these sculptures but also the history of the Geological Court itself. The crux of this issue is whether all the sculptures represent the species *Anoplotherium commune*, or do some represent a second species, '*Anoplotherium gracile*'?

This confusion dates back to the first editions of the Crystal Palace guidebooks (Phillips, 1854) where both species were listed as being present, and has persisted to the modern day: both McCarthy and Gilbert (1994) and Doyle and Robinson (1993) regarded two standing *Anoplotherium* statues

Fig. 4.18 The three Crystal Palace *Anoplotherium commune* in 2015. The positioning of these variably posed individuals alongside the Tidal Lake was especially apt given that, when constructed, *Anoplotherium* was considered a semi-aquatic species.

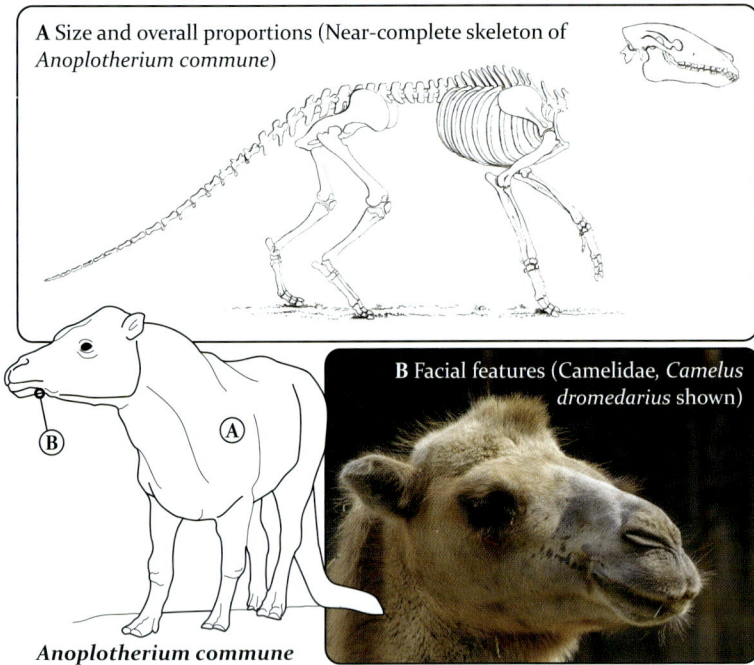

Fig. 4.19 Likely reference fossils and inspirational animal species for *Anoplotherium commune* and *'Anoplotherium gracile'*. Skeletal reconstructions by Cuvier (1822).

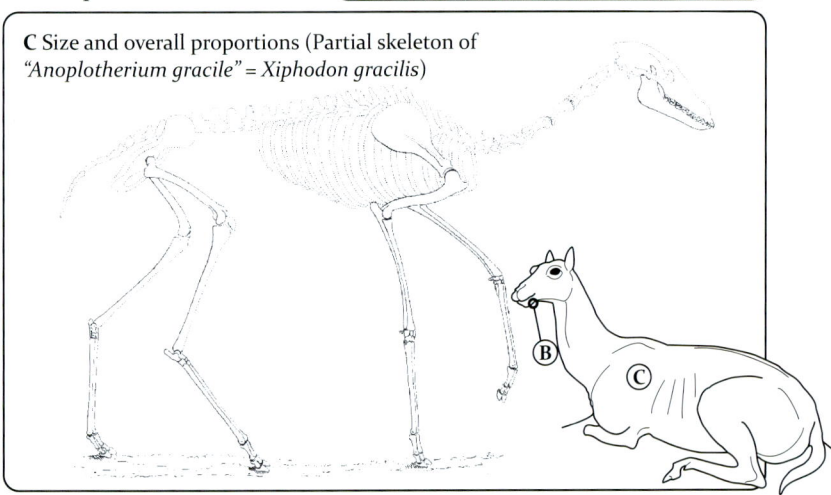

A Size and overall proportions (Near-complete skeleton of *Anoplotherium commune*)

B Facial features (Camelidae, *Camelus dromedarius* shown)

Anoplotherium commune

C Size and overall proportions (Partial skeleton of "*Anoplotherium gracile*" = *Xiphodon gracilis*)

"*Anoplotherium gracile*" (= *Xiphodon gracilis*)

as *A. commune*, and the reposed model as '*A. gracile*'. Given that there are no significant anatomical differences between the three *A. commune* sculptures and the fact that '*A. gracile*' and *A. commune* have always been scientifically recognized as differing markedly in size and shape (*see* Fig. 4.19), this interpretation is almost certainly in error and we can be confident about the *A. commune* identity of the three sculptures at the end of the Tertiary Island.

There is, however, truth to the idea that '*A. gracile*' was present at Crystal Palace. A number of historic guides and an overlooked illustration of Hawkins' workshed show that four '*A. gracile*' sculptures of deer- or llama-like form, and with distinctly camel-like faces, were constructed. Three of these sculptures have vanished with virtually no other historic trace, but one still survives in another part of the park, hiding in plain sight as a misidentified *Megaloceros* fawn. This sculpture has sat next to the *Megaloceros* for an unknown duration with its significance overlooked even by specialists. As discussed below, this finding has important implications for the history of the Geological Court, evidencing more missing models and lost history than was previously realized.

Anoplotherium commune

Three relatively large statues that look like hybrids between deer and big cats are situated along the banks of the Tidal Lake. These 3.6m-long sculptures represent the Eocene European hoofed-mammal *Anoplotherium commune* (*see* Fig. 4.18), perhaps the most obscure animal in the Geological Court for modern visitors but a recognizable species to anyone experienced with palaeontological texts or palaeoart during the 1800s. As a historically important study species of Cuvier, *Anoplotherium* was a mainstay of nineteenth-century palaeontological texts and only disappeared from them at the turn of the twentieth century when more spectacular prehistoric animals took their place (*see* Fig. 2.14). With most of the other Geological Court denizens maintaining their public status in the intervening decades, the *Anoplotherium* represent part of the display that has dated more culturally than scientifically. They are relics of a time when relatively unremarkable, mid-sized extinct mammals were still considered worthy of public attention, and not of specialist interest only.

The modern placement of the *Anoplotherium* is a rough reconstruction of their historic position, the models having been moved in 1953 – along with the *Palaeotherium* – to another part of the park to accommodate a petting zoo. Their historic record is among the poorest of all the sculptures and Hirst Conservation's (2000) paint analysis revealed only tentative indications of their original colour: a straw-like hue seems most likely. All three of the models have required conservation work in major ways, including the replacing of limbs and facial features. One model – the half-crouched individual with a raised head – is a 2001 fibreglass rebuild based on historic images after the original was destroyed.

As with *Palaeotherium*, Hawkins seems to have followed the work of Georges Cuvier closely when restoring *Anoplotherium*, even giving it the same short hair or naked skin Cuvier speculated as being suited to a semi-aquatic habit (Rudwick, 1997). Thanks to Cuvier, *A. commune* was one of the first fossil species to receive dedicated palaeoartistic focus (*see* Fig. 2.14). Two incomplete skeletons recovered from gypsum deposits adjacent to Paris at the turn of the nineteenth century provided grounds to reconstruct most of its osteology between 1804 to 1825 (*see* Rudwick, 1992, 1997 and Hooker, 2007 for discussion and references) and, from these, full skeletal restorations, muscle studies and basic details of life appearance were produced. Some of these were widely reprinted, although Cuvier's muscle studies were not published out of concern that they were too speculative for

Fig. 4.20 Details of the *Anoplotherium* sculptures from 2013 and 2021. Note the different head shapes between the standing and relaxed sitting sculptures: might this reflect depiction of speculative sexual dimorphism?

Chapter 4 – The sculptures: mammals

scientists of the early nineteenth century (Rudwick, 1992, 1997). Today we can recognize Cuvier's unpublished work on *Anoplotherium* as well ahead of its time, foreshadowing practices that would not only be important to palaeoart but also studies of extinct animal biomechanics.

Although Cuvier's work clearly provided the major inspiration for the Crystal Palace *Anoplotherium*, Hawkins deviated from this take by giving *Anoplotherium* very camel-like facial details, including large lips, small, rounded ears and a sloping skull roof (compare Fig. 2.14 with Figs 4.19–20). Referencing camel anatomy was not Hawkins' whimsy but informed by ideas of where *Anoplotherium* sat in mammalian classification. Initially regarded as a creature placed somewhere in the order of life between tapirs and camels (Rudwick, 1997), a camelid affinity has proven to be a long-standing (if not entirely uncontested) hypothesis of *Anoplotherium* affinities (e.g. Prothero and Foss, 2007; Hooker, 2007; Prothero, 2016). In giving *Anoplotherium* a camel-like face, however, Hawkins took his reconstruction away from the true shape of the *Anoplotherium* skull and more robustly inferrable soft tissues. Cuvier's more conservative take, with a lower snout and modest lip tissues, seems more in keeping with *Anoplotherium* cranial osteology and its inferable facial anatomy. This seems to be another example of Hawkins transferring anatomy from living species rather than reconstructing it objectively from fossil bones.

A second deviation, also in error, was Hawkins reconstructing four toes on each foot. *Anoplotherium* feet actually had three toes: two hoofed main digits and somewhat opposable 'thumbs' on the inside of each limb (Hooker, 2007). Cuvier was aware of there being three digits on the forelimbs at least, and it's possible that Hawkins added more toes because he thought the fossils were incomplete or otherwise somehow anomalous. This is a forgivable mistake given that *Anoplotherium* belongs to the even-toed hoofed mammal tribe and, even today, it stands out for its unusual digit configuration.

Other than these relatively minor issues, Hawkins' *Anoplotherium* sculptures have stood the test of time very well, such that some authors still use his 160-year-old models to illustrate *Anoplotherium* today (e.g. Prothero, 2016). Features of note include their strong, flexible-looking tails, the bulging musculature around the top of their limbs, and the robust neck and head features of the standing animal. It is tempting to conclude that the latter, where the head is approximately 11 per cent taller than that of the reclined sculpture, reflects speculative sexual dimorphism, perhaps referencing the comparatively robust skulls of male camels (Yahaya *et al.*, 2012). If so, this must be among the first incorporation of speculative sexual dimorphism in palaeoart (as opposed to the condition in *Megaloceros*, which was well-evidenced in fossils) and was an extremely progressive approach to restoring prehistoric animals in the mid-1800s.

The life-like nature of the three *Anoplotherium* is aided by their varied poses: one stands, one is resting on its belly, and another is in a curious half-crouched pose with an outstretched neck and head, with splaying forelimbs. This individual also has deformed tissues around its neck where long flaps of skin project laterally (this is a genuine feature sculpted by Hawkins – not the result of damage or poor conservation). This pose and the deformed neck skin recall that of a dog shaking itself dry, and it's possible that this behaviour is exactly what is intended to be conveyed. In the early 1800s *Anoplotherium* was regarded as a swimming animal that used its powerful tail to propel itself through water, perhaps like an otter or coypu (e.g. Owen, 1846), and the positioning of this individual close to the waters of the Tidal Lake is consistent with such an interpretation. If correct, this capturing of such a specific behaviour is another extremely progressive approach for nineteenth-century palaeoart – the sort of convention we'd expect in a twenty-first-century restoration, but very unusual against the more generally illustrative or monsterizing attitudes of Victorian palaeoart.

Thanks to the rigorous groundwork laid by Cuvier and Hawkins, our modern takes on *Anoplotherium* are not too different from those of the 1850s. It remains to be seen exactly where *Anoplotherium* fits in hoofed mammal evolution because, other than *Anoplotherium* itself, anoplotheriines are poorly known and we are unable to robustly test their relationships with other species. An affinity with camels remains plausible, but it's also possible that anoplotheriines were part of a European mammal radiation unrelated to any specific hoofed mammal lineage of other continents (Prothero and Foss, 2007). Our interpretations of its anatomy have not changed too much in the last two centuries, but we now realize that *Anoplotherium* was a more robust and barrel-chested species than shown at Crystal Palace (Hooker, 2007). Furthermore, a greater understanding of *Anoplotherium* lifestyles has resulted from analyses of newly recovered fossil skeletons.

It is now suggested that *Anoplotherium* was not a swimming creature, but a fully terrestrial animal adapted for high browsing (Hooker, 2007). Peculiarities of its pelvis, vertebrae, hindlimbs and tail are shared with mammals that

Fig. 4.21 The unusual reconstruction of a half-crouched, half-reared *Anoplotherium* with deformed neck soft tissues recalls the poses used by quadrupedal mammals, such as Beau the greyhound, shaking themselves dry. This may reflect Hawkins' capturing a behaviour associated with a semi-aquatic lifestyle.

Fig. 4.22 *Anoplotherium commune* as reconstructed in 2021. Other than details of face and chest shape, this reconstruction is broadly similar to that seen at Crystal Palace, although we no longer imagine this species as being semi-aquatic.

regularly stand upright on two legs, and it's probable that *Anoplotherium* adopted this pose to browse vegetation from taller bushes and trees (Hooker, 2007). Many deer and antelope are capable of rearing to feed in this way and it's likely that *Anoplotherium* reached 2–3m above the ground at full height – beyond the feeding range of contemporary herbivores other than *Palaeotherium magnum*. The strong, muscular tail, in this hypothesis, becomes a stabilizing appendage rather than a swimming aid, precluding the need for powerful forelimbs and claws for stabilization against tree trunks, as often develop in bipedal feeders (Hooker, 2007). Additional explanations for its bipedality include the capacity to spar with the powerful forelimbs (Hooker, 2007), but *Anoplotherium* should not be imagined as a principally bipedal animal: it likely walked, ran and stood on four limbs most of the time, as depicted at Crystal Palace.

Fig. 4.23 The single surviving '*Anoplotherium gracile*' sculpture, long regarded as a fawn of *Megaloceros*. Both anatomical details of the sculpture and several lines of historic evidence support the re-identification of this sculpture as '*A. gracile*', a species known today as *Xiphodon gracilis*. (2021)

Chapter 4 – The sculptures: mammals

'*Anoplotherium gracile*'

Alongside *Palaeotherium* and *Anoplotherium commune*, '*Anoplotherium gracile*', known today as *Xiphodon gracilis*, was another one of Georges Cuvier's important and influential studies of fossil mammals from the Paris Basin (see Fig. 2.14). It joined these species in early palaeoartistic depictions and would have been well known to any students of fossil life in the early 1850s. Such status must have seen '*A. gracile*' rank high on Hawkins' list of subjects for the Geological Court, and yet traditional interpretations of the park suggest either no statues of this species were created or, alternatively, that one of the *A. commune* sculptures must represent '*A. gracile*' (Doyle and Robinson, 1993; McCarthy and Gilbert, 1994). The latter is clearly not the case. '*A. gracile*' was a small, long-legged, delicate creature resembling a small deer or camel, and thus entirely different from the robust, pony-sized *A. commune*. This distinction was so well reflected in literature and art even before the Crystal Palace opened that '*A. gracile*' was reclassified as *Xiphodon* in 1845. Hawkins' apparent refusal or ignorance of this name change may explain some of the uncertainty around the identity of the *Anoplotherium* sculptures: he continued calling *Xiphodon* '*Anoplotherium gracile*' long after other scholars – including Owen (1857) – embraced its new name.

The most likely resolution to this confusion is actually nothing to do with the *Anoplotherium* statues at all. A variety of sources strongly imply that four genuine *Xiphodon* sculptures were built and placed in the park and, moreover, that one survives today as a misidentified *Megaloceros* fawn (see Fig. 4.23). It is only in the process of compiling this book that the probable truth behind the Crystal Palace *Xiphodon* has been pieced together and much about their history still remains to be investigated. If correct, this assertion implies another major gap in our records of the Geological Court, necessitating that three additional *Xiphodon* sculptures once existed and then disappeared; that the sole remaining example was removed from its original site; and that a llama-like animal could be misidentified as a juvenile deer by several generations of experts. Though seemingly far-fetched, these extraordinary implications are the best explanation for a number of facts and circumstances surrounding the alleged 'fawn' sculpture and related historic documents.

A close look at the anatomy of the *Xiphodon* sculpture is the first indication that its placement among the *Megaloceros* is an error. It is a small (*c.* 1.7m long from snout to tail-tip), long-necked hoofed mammal with a small head, large eyes, long, robust body and a narrow, tapering snout. Although its proportions are superficially consistent with what could be expected of a juvenile deer, it also has a flattened, camel-like nose and bifurcated upper lip, as well as a long, slender tail. These features are not mirrored in the relatively robust, short-tailed, large-nosed and wide-muzzled *Megaloceros*. Indeed, the thin *Xiphodon* tail is actually the same length as the wider tails of the Giant Deer, despite their contrasting body sizes. Save for a small patch on the chest, the fur of *Xiphodon* is also almost entirely depicted as short, further contrasting with the maned *Megaloceros* sculptures. But where *Xiphodon* differs from the deer it resembles Hawkins' *Anoplotherium commune*, which also has camel-like nostrils and lips, short hair and a long tail. This is surely no coincidence, given that – like *Anoplotherium* – *Xiphodon* was also thought to have affinities to tapirs and camels in the early 1800s (Rudwick, 1997).

The Crystal Palace *Xiphodon* is also inescapably similar to Georges Cuvier's long-legged, long-necked reconstructions of '*A. gracile*' (see Fig. 4.19), which were surely Hawkins' anatomical references for this species. Indeed, the only major inconsistency between the sculpture and Cuvier's work is that Hawkins added small digits above the principal hooves, a mistake he also made with *A. commune*.

In addition to these anatomical points, several historic documents corroborate the existence of *Xiphodon* in the Geological Court. The most informative is an 1854 image of Hawkins' worksed showing four llama-like sculptures where two stand, one is fully reposed, and another (leftmost in the image) is the sitting *Xiphodon* we know today (see Figs 3.21B, 4.24). There is little question that this image captured a real scene of Hawkins' workshop as it includes precise depictions of several recognizable statues (two *Megaloceros*, one '*Labyrinthodon pachygnathus*' and the sitting *Xiphodon*), so the three unfamiliar sculptures are surely real objects drawn by an eyewitness, not fanciful creations. This illustration is our best insight into the appearance of the entire *Xiphodon* group and suggests that their arrangement was consistent with the rest of the Geological Court: a scene of calm, relaxed animals.

Written works corroborate their existence, too. Several articles and guidebooks mention four antelope- or camel-like sculptures that, as described, do not match the other species depicted in the park. McDermott's (1854) *Routledge's Guide* mentions 'four graceful llama-like' *Anoplotherium* sculptures (p. 192) and an 1854 *Hogg's Instructor* article describes them as 'like a number of antelopes or deer' (p. 285). Official park guides, including corrected and updated versions released after the park opened, also continually mention

Fig. 4.24 Detail of an 1854 workshed illustration showing the four *Xiphodon* sculptures, the left-most of which is clearly the '*Megaloceros* fawn' that survives in the park today. The fate of the others is unknown. Note the individual lying on its side, as is typical of guanacos, a likely analogue species for this reconstruction.

the presence of '*Anoplotherium gracile*' (e.g. Phillips, 1855, 1856), with one – the anonymous 1871 edition – giving their location as 'under a tree'. In none of these works is any mention made of a *Megaloceros* fawn. These persistent reports, the 1854 illustration, and the anatomical distinctions of the '*Megaloceros* fawn' sculpture are difficult to explain as anything other than evidencing a series of largely overlooked *Xiphodon* sculptures.

Of course, for *Xiphodon* to have been part of the Geological Court menagerie three sculptures must have escaped all but cursory historic records and also vanished without any physical trace. This may seem unlikely, but the incontrovertible and barely documented relocations and disappearances of other mammal models from the Geological Court set good precedent for such actions. There are, sadly, sufficiently large gaps in our knowledge of the Crystal Palace Dinosaurs that three full-size *Xiphodon* sculptures could have vanished under a number of plausible scenarios. The disruption of the Tertiary Island species at hands of park custodians (*see* Chapter 12), their history of questionable conservation, and well-documented vandalism of the Geological Court offer plenty of scope for the removal and probable destruction of three additional sculptures, as well as the misidentification and relocation of the fourth. Furthermore, as a Paris Basin species, *Xiphodon* would have originally been placed close to *Palaeotherium* and *A. commune* (*see* Figs 1.4 and 12.7), and thus in the region affected by zoo developments in the mid-twentieth century, an event that seems to have caused much disorder for the Tertiary Island sculptures. However the various *Xiphodon* sculptures met their respective fates, the poor documentation and low interest in the Crystal Palace mammals have allowed their existence to pass unnoticed, and for the erroneous '*Megaloceros* fawn' identification to become established for the surviving statue.

Fig. 4.25 Details of the Crystal Palace *Xiphodon gracilis* as seen in 2013 and 2021.

Our poor records of the Geological Court *Xiphodon* limit our capacity to interpret the palaeoartistry of Hawkins' '*A gracile*' models. The 1854 illustration and the surviving sculpture suggest they were authentic takes on the anatomy of *Xiphodon* as known in the mid-1800s, and closely based on Cuvier's pioneering skeletal studies and concepts of anoplothere affinities. It seems all four sculptures had the same body plan without, as is seen in some of the other mammals, obvious sexual dimorphism. Their bodies look well-muscled but they were not cast as especially bulky or heavyset creatures, meeting early interpretations of this species as a fast runner (Owen, 1846; Rudwick, 1997). The camelid anatomical influence is even more obvious than with *Anoplotherium commune* and the overall body plan is very reminiscent of the South American guanaco *Lama guanicoe*.

Fig. 4.26 *Xiphodon gracilis* as we might reconstruct it today: essentially the same as reconstructed by Hawkins, as a small, slender, deer-like species.

The herding behaviour of these South American llamas may have influenced Hawkins' decision to restore a group of '*A. gracile*', and they were probably arranged in a naturalistic group like that of *A. commune*. Given the svelte bodies of these animals, all four sculptures may have been cast primarily in metal, as is the case of the surviving example. Being constructed of such materials has meant the remaining *Xiphodon* has avoided major conservation needs but its ears have proven highly vulnerable to damage. The 1854 illustration shows that these were originally narrower and more pointed than the rounder, fawn-like ears of recent restorations. Paint analyses have not been conducted on this statue on the grounds that, when such studies were done, its distinctiveness was not recognized but, unsurprisingly, photographs show that *Xiphodon* has been painted with the same browns and creams as the *Megaloceros* for much of its recent history.

As with most of Hawkins' mammal sculptures, his '*A. gracile*' work has stood the test of time. Today, we would reconstruct this species broadly as he did, with our main distinctions incurred through modern ideas of where *Xiphodon* fits among mammalian evolution. The relationships of *Xiphodon* and its kin, the Xiphondontidae, are only provisionally explored but they seem to be a small group of hoofed mammals that – like the anoplotherids – were restricted to the Palaeogene of western Europe. Also like *Anoplotherium*, some doubt now exists over the camelid affinities of xiphodontids. A camel link remains promoted by some, but other workers regard xiphodontids as close relatives of the ruminants – cows, deer, giraffes, sheep and so on – that converged on camel-like anatomy (*see* discussion in Prothero and Foss, 2007). A modern reconstruction of *Xiphodon* may accordingly dial back their camelid-like features in favour of more ruminant-like facial tissues.

The specific lifestyles of xiphodontids also remain to be investigated in detail, but their dentition is broadly comparable to that of ruminant mammals, and thus indicative of a diet of tough vegetation. Little doubt can be held about their likely capacity for speedy locomotion: their long limbs with especially elongated distal regions and reduced digit counts are classic examples of running adaptations in hoofed mammals. Sadly, given their importance to early palaeontology, *Xiphodon* has largely been overlooked by the last century of palaeoartistry and few contemporary reconstructions exist. This makes it all the sadder that Hawkins' takes on these animals have been largely lost and that the only surviving example has been overlooked for so long.

CHAPTER 5

The sculptures: *Mosasaurus hoffmanni*

Visitors approaching the northern extent of the Secondary Island are welcomed to the next section of the Geological Court by a large, partial restoration of the giant mosasaur *Mosasaurus hoffmanni*. This Late Cretaceous marine reptile is depicted emerging from the water close to the pterosaur-perching Chalk cliff on the island's northeast coast, with the head, back and right forelimb visible. The gaping mosasaur is the northern gatekeeper of the Age of Reptiles, its huge mouth with its two rows of teeth promising a different experience to the sedate mammals of the preceding island. It faces the weir that runs between the Tertiary and Secondary islands, the waterway positioned to emphasize the transition between their representative epochs (even if, as discussed in Chapter 3, the simulated geology presented a more complex arrangement). Until 2017 the weir supported a peninsular walkway and a gated bridge that permitted controlled access to the Secondary Island. Access has now been shifted to the southwest end (*see* Chapter 13) and the removal of the gate has improved views of the mosasaur from the Tertiary Island.

At first glance it may seem that the *Mosasaurus* sculpture is partially hidden, perhaps awaiting full exposure from the long-defunct Tidal Lake system like the other marine reptiles of the Geological Court. Instead, the *Mosasaurus* was deliberately left incomplete: one of only two restorations in the entire Crystal Palace fauna to be built in this manner. Approximately 6.7m of the animal has been created, accounting for much of the body length estimated for this animal in the mid-1800s (Mantell, 1851; Owen, 1854). The sculpture has received several rounds of conservation to its teeth, jaw tips, flipper and skin, all of which have cracked, broken or fallen off on several occasions. Moisture-accelerated decay of paint layers has removed all records of the original colour of the *Mosasaurus* and archived colour details in the forms of photographs and film footage are hard to come by on account of its recessed location. It seems to have been various shades of green since at least the 1950s, however.

Unlike the other marine reptiles of the Geological Court, mosasaurs were not common palaeoart subjects in the early to mid-1800s. This is curious in light of mosasaur remains – including skulls, vertebrae and limb elements – being known from Chalk deposits of Europe and North America since the mid-1700s. Indeed, the discovery of two *Mosasaurus* skulls in the Netherlands during the 1760s and 1780s are famous historic events: one specimen was even seized by the French army in 1894 during the French Revolutionary Wars, ostensibly for scientific study (a story recounted by Owen in the 1854 Geological Court guide). These remains were among the first giant extinct reptiles known to science and they caused much stir among scholars of the time. Initially regarded as crocodiles or whales, the true identity

Fig. 5.1 The Crystal Palace *Mosasaurus hoffmanni*, one of only two deliberately incomplete sculptures in the Geological Court. The level of the Tidal Lake is unusually low in this image, showing the full extent of the sculpture and concrete foundation. (2018)

of *Mosasaurus* as a type of marine lizard, closely related to snakes and monitor lizards, was established at the turn of the nineteenth century. Other mosasaur remains were found in the USA around the same time, including complete and articulated skulls recovered in the 1830s.

Although not as well represented as ichthyosaurs or plesiosaurs, *Mosasaurus* was thus as well known to scientists and artists of the mid-nineteenth century as the first dinosaurs, but life reconstructions from this time are scant. Perhaps, as suggested for the similarly artistically neglected *Megaloceros*, mosasaurs represented older discoveries of the eighteenth century and were simply not as exciting to palaeoartists of the early 1800s? Whatever the reason, the Crystal Palace mosasaur is among the first, if not *the* first, legitimate life restoration of a mosasaur – created almost ninety years after their fossils were found. (The 'legitimate' comment here deserves explanation: Felix Guérin's 1834 *Dictionnaire pittoresque d'histoire naturelle et des phénomènes de la nature* features an illustration labelled as a giant fossil crocodile from Maastricht, but it's a copy of *Megalosaurus* artwork from an older illustration.)

Mosasaurus was still imperfectly known in the mid-1800s, with the most notable specimens being near complete skulls from Chalk deposits in the Netherlands and the USA (*see* Figs 5.2–5.3). This lack of material has been used to explain why the Crystal Palace *Mosasaurus* was only partially restored. Owen (1854) states:

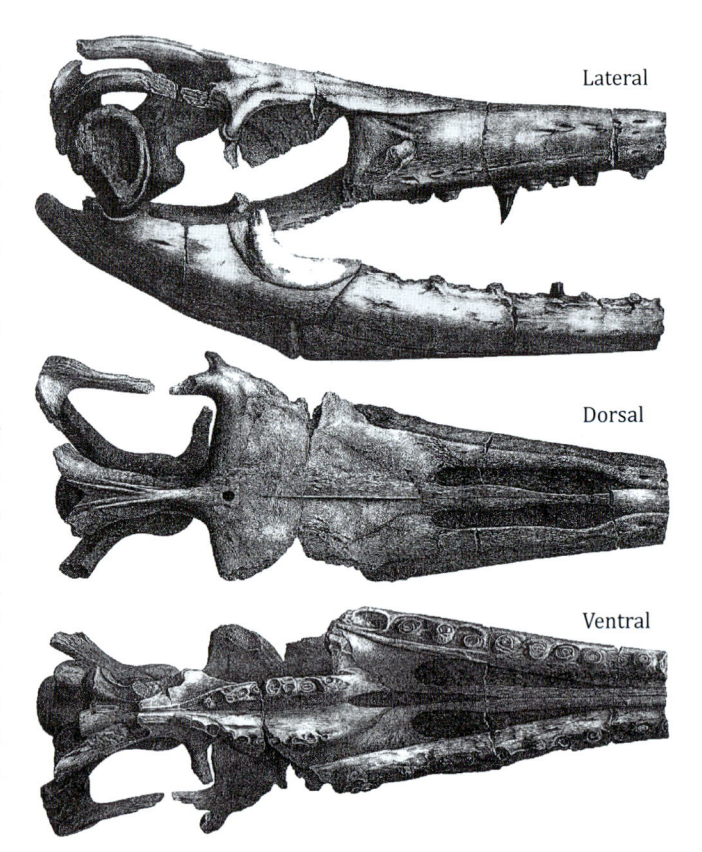

Fig. 5.2 One of the earliest articulated skulls of a mosasaur: the cranial remains of *Mosasaurus missouriensis* from South Dakota in the early 1800s. First described by Richard Harlan in 1834 and then Georg August Goldfuss in 1845, this specimen showed many realities of mosasaur faces overlooked by Hawkins and Owen. The result was a less-than-accurate *Mosasaurus* reconstruction at Crystal Palace.

Fig. 5.3 Fossil material and modern species likely influencing Hawkins' *Mosasaurus* sculpture. *Mosasaurus* skull from Cuvier (1812).

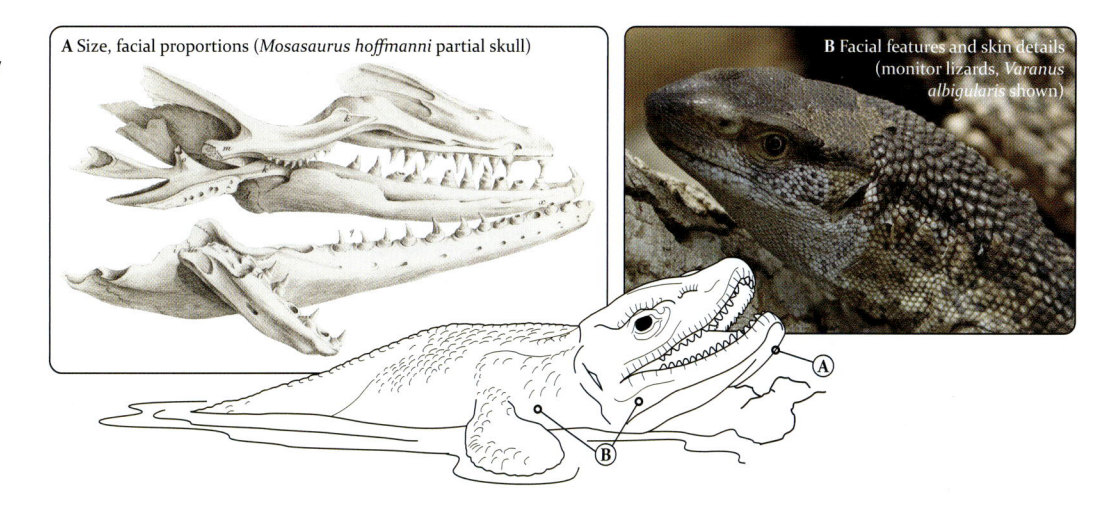

A Size, facial proportions (*Mosasaurus hoffmanni* partial skull)

B Facial features and skin details (monitor lizards, *Varanus albigularis* shown)

Of this animal almost the entire skull has been discovered, but not sufficient of the rest of the skeleton to guide a complete restoration of the animal. The head only, therefore, is shown, of the natural size…

OWEN, 1854, P.10

This justification is at odds with several other historic facts, however. Firstly, contrary to Owen's comment, sufficient material of *Mosasaurus* had been recovered by the 1850s for a basic overview of its anatomy to emerge. It was clearly a very large reptile (estimated at 24ft [7.3m] long by Mantell in 1851 and 30ft [9.2m] by Owen in 1854) that, while possibly amphibious and capable of skink-like locomotion on land (Goldfuss, 1845), was certainly a powerful swimmer. By 1851 Mantell was able to provide a loose but essentially correct description of *Mosasaurus* form and lifestyle:

> …the entire length of the skeleton is estimated at twenty-four feet; thus the head is nearly one-sixth of the total length… The tail is only ten feet long, and therefore but half that of the total length… The bones of the extremities are but imperfectly known; those attributed to the *Mosasaurus* are said to indicate members adapted for natation rather than for progression on land, and to support the inference of M. Cuvier, that the original was a marine animal of great strength and activity, having a large vertically expanded tail, capable of being moved laterally with such force as to constitute a powerful instrument of progression, capable of stemming the most agitated waters.
>
> MANTELL, 1851, P. 199

This grade of understanding is comparable to that achieved for several other Geological Court species and it is not clear why Owen would single out this creature as being too poorly known for a complete reconstruction (we can especially look at *Dicynodon* as a comparator, this genus being even more poorly known than *Mosasaurus* – see Chapter 11). Might other factors have influenced the decision to only partially build *Mosasaurus*? A full restoration of this animal would have been very large and, if complete, the tail of this half-submerged sculpture may have extended so far from visitors as to have been barely visible. This may have seemed like a near-pointless addition for the tremendous work it would have entailed. Alternatively, perhaps the additional time, monetary and space demands of another giant build were beyond the already stretched resources of the Geological Court team?

As potentially the first restoration of a mosasaur, Hawkins' sculpture is a reasonable one. The head is wide and boxy, consistent with Owen's (1854) interpretations of the skull being 2.5 × 5 feet (0.76 × 1.5m). Owen's length estimate is approximately accurate for a large *Mosasaurus* skull (Lingham-Soliar, 1995) but is much too wide (*see* below). In fairness to Owen, the Dutch *Mosasaurus* skulls he was familiar with were incomplete and disarticulated, and at face value seem to represent broad-faced animals. The short neck and long torso are apt, and the mostly speculative forelimb is of a length and breadth roughly in line with real mosasaur anatomy. The body is muscular and only lightly contoured, as we might expect for a powerful swimmer. As with '*Labyrinthodon*', Hawkins accurately captured both the robust principal dentition as well the secondary, palatal tooth rows hiding within the mouth – features hidden from most visitors but obvious at close inspection.

Fig. 5.4 Details of the Crystal Palace *Mosasaurus* head, mouth and body. (2018)

Much of the facial anatomy seems to have been modelled on that of monitor lizards, especially short-faced, boxy-headed species of the *Polydaedalus* subgenus, such as savannah and rock monitors (*Varanus exanthematicus* and *albigularis*, respectively; see Fig. 5.3). Extensive lips, voluminous tissues around the eye socket, vertically-alligned, membrane-sealed ear openings, and laterally facing, posteriorly positioned nostrils are among the more obvious monitor facial features. The skin texture – a mix of larger, polygonal scales surrounded by smaller ossicles – also recalls monitor anatomy, especially the osteoderm-studded skin of certain large and robust species, including komodo dragons (*V. komodoensis*) and rock monitors. This referencing of *Varanus* was entirely sensible given the close affinities between mosasaurs and these lizards, and we still use these lizards as reference species when restoring mosasaurs today.

In all, Hawkins' take on *M. hoffmanni* was a valiant first effort at restoring this giant marine lizard from scant material, but it has not dated as well as his other marine reptiles. His interpretation is almost idiosyncratic to the Geological Court, save for an 1860s artwork by Louis Figuier which seems to have referenced his sculpture (see Fig. 5.5). Ultimately, the discovery of many superior American mosasaur fossils within years of Crystal Palace Park opening exposed Hawkins' work as a flawed take on mosasaur anatomy. Indeed, one American mosasaur fossil showed that the Crystal Palace reconstruction was problematic even when first designed and built. This near complete, articulated and undistorted skull of '*Mosasaurus maximiliani*' (now identified as *M. missouriensis*), reported from the midwestern United States in 1845 by the German palaeontologist Georg August Goldfuss, showed that mosasaurs had a much narrower cranial profile than assumed by Hawkins and Owen, and also that the nasal openings were on the upper surface of the snout (see Fig. 5.2).

It's likely that this find was simply overlooked, rather than deliberately ignored by the Crystal Palace team, and they were not the last to miss this pioneering work. Goldfuss' *Mosasaurus* skull description was passed over for much of the nineteenth century such that many of his observations were effectively repeated by American palaeontologists such as Joseph Leidy, Edward D. Cope and Othniel Marsh in the 1860s and 1870s (Williston, 1914). It is curious, given the incredible enthusiasm to understand giant extinct reptiles at this time, that such an important discovery and interpretation were overlooked for so long.

Other reconstruction errors were more forgivable and relate to Hawkins not outfitting his mosasaur with sufficient adaptations to a marine existence. Although satisfied that *Mosasaurus* was primarily a swimming animal, scholars of the mid-1800s had not yet ruled out some degree of amphibious locomotion (Goldfuss, 1845), perhaps explaining why Hawkins' sculpture has coarse skin and laterally facing nostrils, and looks better suited to a crocodile-like

Fig. 5.5 A rare European mid-nineteenth-century take on *Mosasaurus*, by Edward Riou (from Louis Figuier's 1863 book, *La Terre avant le déluge*). Relegated here to a supporting role in a scene focused more on botany and invertebrates, this tableau exemplifies European palaeoartistic disinterest in mosasaurs. It was not until American palaeoart became established, from the late 1860s onwards, that mosasaurs became popular restoration subjects. Riou's *Mosasaurus* has a head shape seemingly referencing the Crystal Palace model, showing the reach and influence of Hawkins' work.

amphibious life than a fully marine one. Today, however, we regard mosasaurs as animals highly adapted for life at sea: the lizard equivalent of a shark or toothed whale. Remains of their body outlines and skin show that their flippers were tapered into efficient hydrofoils (Lindgren *et al.*, 2013) and that their skin was covered in tiny, interlocking scales with medial ridges aligned with the long axis of the body, almost reminiscent of shark skin (Lindgren *et al.*, 2011). Their bodies were stiffened and their tails adorned with prominent fins, recalling those of fish and ichthyosaurs in development (Lindgren *et al.*, 2013).

Mosasaurus hoffmanni remains a relatively poorly known species (Street and Caldwell, 2017) but it represents one of the last and most marine-adapted lineages of the group, as well as one of the largest species – perhaps reaching 10 or 11m long, based on better known close relatives. Some anatomical issues aside, the impression given by Hawkins' sculpture of *Mosasaurus* being a gigantic, powerful swimming carnivore remains an apt ecological interpretation of this species: an apex predator widely distributed across the early Atlantic Ocean in the Late Cretaceous.

Fig. 5.6 *Mosasaurus hoffmanni* as we might restore it in 2021. Our understanding of mosasaur life appearance has changed in recent years: rather than crocodile-like swimming lizards, we now view them as highly adapted to marine life, and thus convergent with whales, sharks and other large pelagic fish in many details of life appearance.

CHAPTER 6

The sculptures: flying reptiles

The waters separating the Secondary and Tertiary Islands are overlooked by two reptilian sentinels perched atop a tall Chalk cliff. These are the surviving pterosaurs, or flying reptiles, of the Geological Court. Four pterosaur models were originally built: the two large individuals situated in the Cretaceous Chalk region alongside the *Mosasaurus* (see Fig. 6.1), and two smaller, Jurassic specimens between the *Teleosaurus* and *Megalosaurus* (see Fig. 6.2). The original small pterosaurs disappeared at some point in the 1930s (McCarthy and Gilbert, 1994) and were replaced in 2001 with fibreglass replicas as part of large-scale renovation of the Geological Court. Alas, these suffered the same fate as their forebears, only standing until 2005 before they were broken and removed from the Geological Court. The Friends of Crystal Palace Dinosaurs tracked them down in 2014, finding that the shattered remains of these second generation Jurassic pterosaurs had been thrown, entirely unrecorded, into an equipment shed. The larger models have not escaped misfortune either, their relative delicacy making them vulnerable to deterioration.

Pterosaurs were well-established components of paleontological research by the 1850s as well as regular fixtures in palaeoart (see Figs 1.8–1.9, 6.3). Indeed, the oldest known scientific life restorations of prehistoric animals are sketches of the first-known pterosaur fossil, *Pterodactylus antiquus*,

Fig. 6.1 *'Pterodactylus' cuvieri*, today known as *Cimoliopterus cuvieri*, atop their Chalk cliff at the apex of the Secondary Island in 2009. Four pterosaurs were built for the Geological Court, of which only these two remain.

from 1800 (Taquet and Padian, 2004). Pterosaurs are considered challenging animals to interpret from fossils because their anatomy was strongly modified for powered flight and thus radically different from that of other reptiles. Their wings consisted of long, hollow bones and a highly unusual hand where the fourth (or 'ring') finger was tremendously elongated to support a flight membrane. Their proportions were among the most extreme of any limbed vertebrate, with their inflated heads and arms appearing outlandish against their smaller bodies and legs.

Fig. 6.2 Two generations of Oolite pterosaurs, '*Rhamphocephalus bucklandi*'. **A)** the original Oolite pterosaur sculptures photographed before their disappearance in 1930; **B–D)** details of the fibreglass replacements installed in 2002; **E–F)** the fate of the 2002 fibreglass replacements: broken by strong winds or vandals (or both) in 2005, their broken bodies are now in storage.

Hawkins' pterosaurs were built, like the mammal sculptures, around iron frames that anchored them to rocky foundations. According to an April 1855 *Observer* article, the wings were originally made of leather sown double around the iron framework: these have since been replaced with sturdier material matching the approximate shape of the originals. The same article indicates that one of the large pterosaurs was the final palaeontological model added to the Court before work ceased in autumn 1855. The four pterosaur sculptures were constructed as crouched on all fours or standing upright with outstretched wings, but a journalist for *Hogg's Instructor* reported that more dynamic posing was initially planned:

Mr Hawkins has bestowed great pains upon his restorations of these creatures and aims to represent them so as to exhibit the various powers they possessed. One is seen as if about to plunge into the water; another running like a bird upon the ground; a third suspending itself by its claws from the face of a rugged cliff; while a fourth is seen crouching with its wings spread, as if about to mount into the air.

ANONYMOUS, 1854, P. 284

The history of the Crystal Palace pterosaurs is rather patchy. Our colour data is relatively good: Hirst Conservation's (2000) paint analysis suggests that the large pterosaurs were originally covered with a golden/green glaze over a white base, but they were later painted grey. Most recently they

Fig. 6.3 Pterosaurs were popular subjects of early palaeoartists and featured in several famous works, including in **A)** John Martin's *The Sea Dragons as They Lived* (1840); **B)** as cliff-climbers by Georg Goldfuss (1831); **C)** flying above the packed Jurassic sea in De La Beche's *Duria Antiquior* (1830), with one being seized by a *Plesiosaurus*.

have been painted with a creamy-yellow hue. Records of these changing colours are incomplete because the maturing foliage of the Secondary Island has often obscured the Chalk pterosaurs in archival photographs, and the same issue prohibits knowledge of their original head shapes, which have been replaced at least twice in different rounds of conservation. The wings are also prone to damage and have been repaired several times, although a secondary set of wing membranes situated behind the legs have largely or entirely decayed without recent replacements. The details of the smaller, now missing models are also only partially understood thanks to poor documentation before their disappearance ninety years ago. Most photos of the original pterosaur sculptures are imperfect in one way or another so any good-quality images you see of these statues – whether the large or small species – include restored or replacement components to greater or lesser extents.

Hawkins' pterosaur restorations are part of two centuries' worth of scientific and artistic efforts to understand the anatomy and appearance of these challenging fossil subjects. Today, we regard them as elegant, sophisticated creatures that are sadly lost to time, but scholars of the early nineteenth century often compared them to monsters: adaptable but grim species that flew, crawled and swam around primordial Earth. This quality was even put forward in promotional articles about the Geological Court as well as in Owen's 1854 guidebook, both of which repeated comparisons of pterosaurs to the Fiend from Milton's *Paradise Lost* by the geologist William Buckland:

> The Fiend,
> O'er bog, or steep, through strait, rough, dense, or rare,
> With head, hands, wings, or feet, pursues his way,
> And swims, or sinks, or wades, or creeps, or flies
>
> MILTON, *PARADISE LOST* BOOK II
> (QUOTED IN OWEN, 1854, P. 13)

The Crystal Palace pterosaurs were among the first attempts to model these animals in three dimensions but, for this novelty, are fairly typical interpretations of pterosaur life appearance for the mid-1800s. They borrowed heavily from the Late Jurassic German pterosaur *Pterodactylus antiquus*, most obviously with their long necks and jaws with peg-like teeth. It was inevitable that Hawkins' work would lean on *Pterodactylus* to a great extent as this was one of the only completely known pterosaur species of the time as well as the assumed generic identity of the species modelled for the

98 Chapter 6 – The sculptures: flying reptiles

Fig. 6.4 Probable fossil and living animal references in the Crystal Palace pterosaurs. *Pterodactylus* skeleton from Cuvier (1812), Stonesfield Slate bones from Huxley (1859).

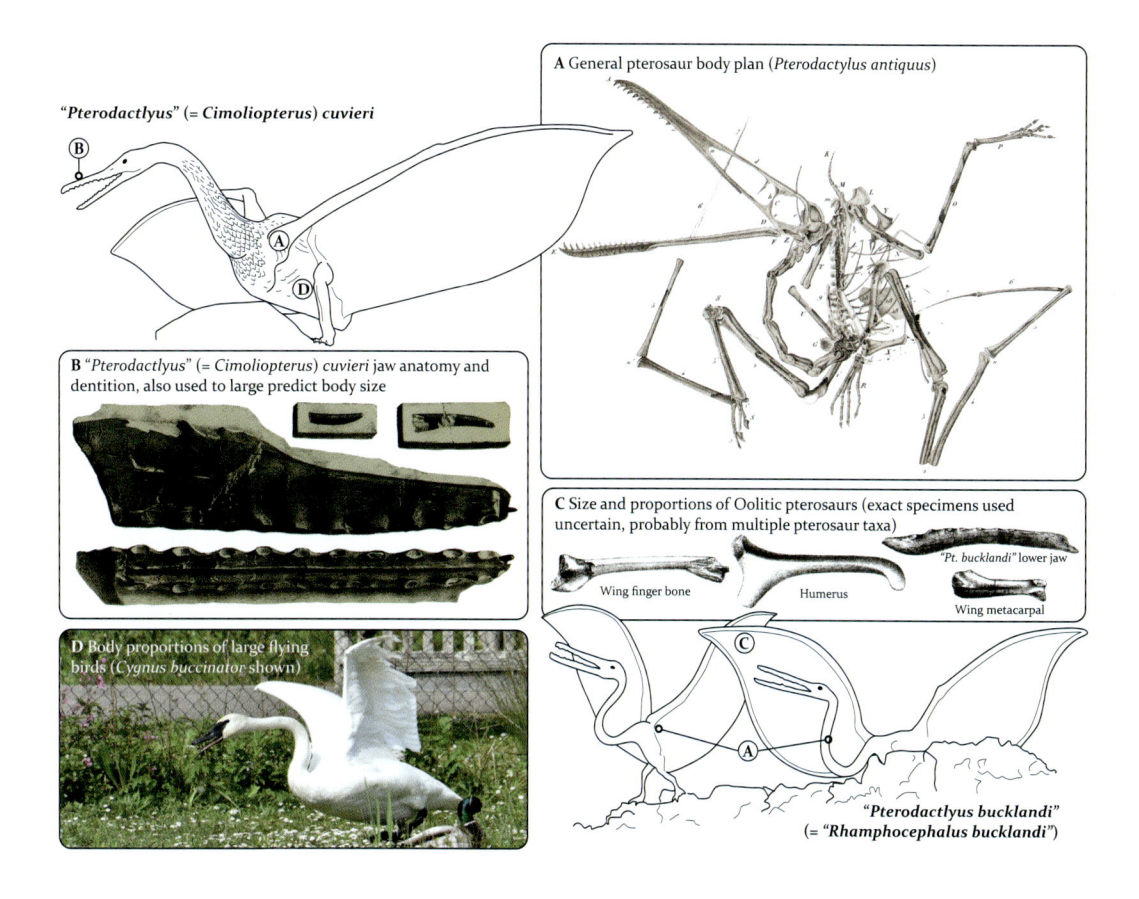

A General pterosaur body plan (*Pterodactylus antiquus*)

B "*Pterodactlyus*" (= *Cimoliopterus*) *cuvieri* jaw anatomy and dentition, also used to large predict body size

"*Pterodactlyus*" (= *Cimoliopterus*) *cuvieri*

C Size and proportions of Oolitic pterosaurs (exact specimens used uncertain, probably from multiple pterosaur taxa)

Wing finger bone — Humerus — "*Pt. bucklandi*" lower jaw — Wing metacarpal

D Body proportions of large flying birds (*Cygnus buccinator* shown)

"*Pterodactlyus bucklandi*" (= "*Rhamphocephalus bucklandi*")

park. In reality, both Crystal Palace pterosaur species were very different from *Pterodactylus*, but Victorian scientists had yet to fathom quite how diverse pterosaurs were and assigned virtually all their fossils to the genus *Pterodactylus*.

In doing so, they created a wastebasket taxon: a collection of fossils united through gross anatomical similarity, but actually representing diverse genera and species. This practice created an overly simplistic, homogenized view of pterosaur anatomy that was reflected in artwork of the time, as well as probably originating use of the term 'pterodactyl' as a common way to refer to all flying reptiles. This is not a vernacular term widely used among palaeontologists, however, as names like *Pterodactylus*, pterodactylid and pterodactyloid have narrower, more specific taxonomic definitions within the pterosaur group.

A defence of these overzealous taxonomic acts is that Victorian researchers and early palaeoartists had very little substantial pterosaur material to work with. A few small, well preserved pterosaurs from Germany and, to a lesser extent, the southern UK were the only flying reptile fossils yet discovered which were not isolated or broken bones (Wellnhofer, 1991; Witton, 2013). The species reconstructed by Hawkins, '*Pterodactylus*" *cuvieri* and '*Pterodactylus bucklandi*' were, and still are, represented from highly fragmentary remains that are not anatomically informative without additional specimens for comparison. Given this state of knowledge, Hawkins could do little more than follow the body plan suggested by *Pterodactylus*. It was not until the second half of the 1800s that fossil discoveries allowed artists to start showing more diversity of pterosaur body shapes, principally through illustration of the long-tailed *Rhamphorhynchus* and *Dimorphodon*, as well as the crested, gigantic *Pteranodon*.

Despite the imperfect reference material Hawkins had to hand, his pterosaur sculptures captured the principal anatomical details of flying reptiles well, although they differ from modern reconstructions in several ways. Along with errors in digit counts, they are covered in finely sculpted scales, a restorative option highlighted with confidence in Owen's (1854) guidebook. This reconstruction choice was typical for early pterosaur art, but ignored data presented in 1831 by German palaeontologist Georg August Goldfuss, demonstrating that pterosaurs had a fuzzy, hair-like covering. This is a second case of Hawkins and Owen overlooking Goldfuss' observations (*see* Chapter 5) and, once again, their work has dated more for it. Hawkins and Owen were not alone in ignoring Goldfuss' fuzzy pterosaurs, however, and it was not until the 1970s that pterosaur skin fibres would be an accepted fact of their palaeobiology. Indeed, it is only

Chapter 6 – The sculptures: flying reptiles

recently that Goldfuss' 1830s work has been fully vindicated, where modern fossil imaging techniques have proved beyond all doubt that his soft-tissue interpretations were accurate (Jäger et al., 2018).

A second major difference is that Hawkins seems to have left the legs of his pterosaurs free of the principal wing membranes, instead anchoring the wing tissues to the body alone. This, as well as the oval or elliptical shape he bestowed on the membranes themselves, creates the impression of a bird-like wing arrangement. Such reconstructions were not uncommon in early pterosaur art but it was not the only restoration option available to the Crystal Palace team: many contemporary artworks showed a bat-like membrane arrangement where the wing tissues stretched onto the hindlimb (see Fig. 6.3B–C). Hawkins cannot be accused of making an obvious error with this decision, however. The extent and attachment sites of pterosaur wing membranes have remained controversial among pterosaur researchers for decades and numerous conflicting models of wing shape have been proposed in the last two centuries. It is only in recent years that sufficient fossil data has accumulated to show that hindlimb wing membrane attachments are most likely (Elgin et al., 2011), although even this conclusion remains contested.

'Pterodactylus' cuvieri

The only pterosaurs encountered by visitors to the Geological Court today are large (5m-wingspan, 1.6m-tall) sculptures representing species from Cretaceous Chalk deposits of the UK (see Figs 6.1, 6.5). At least one of these was identified by Owen (1854) as *'Pterodactylus' cuvieri*, a large pterosaur known from a partial and upper jaw first described by James Scott Bowerbank in 1852. It's not clear if both models represent this species, but they are similar enough in size and anatomy to assume this is the case. Although somewhat difficult to see today, especially when the surrounding vegetation is in full leaf, the Chalk pterosaurs would have been obvious and imposing denizens of the Geological Court in its original, less forested state.

Today, *'Pterodactylus' cuvieri* is recognized as its own genus, *Cimoliopterus*. This is a significant animal for a number of reasons: as well as being among the last of the toothed pterosaurs, *Cimoliopterus* was among the first pterosaurs known to have possessed a bony cranial crest – an extension of the skull tissues common to many pterosaurs and likely used in visual communication within species. This element is incompletely preserved in *Cimoliopterus* fossils but is prominent enough that a low, humped crest at the tip of the upper jaw cannot be doubted. This component was not factored into the Crystal Palace *Cimoliopterus*, however, it seemingly being assumed that the jaw tips of this animal were more akin to those of *Pterodactylus*.

The Geological Court *Cimoliopterus* sculptures have the distinction of being the oldest life reconstructions of 'giant' flying reptiles. The first pterosaurs known to palaeontologists were within the size range of living birds (i.e. wingspans of 3m or less), but Bowerbank's Chalk species – which he estimated as being 16.5ft (5.2m) across the wings, and Owen estimated at a similar 18ft (5.5m) – was clearly much larger. Although a 5–6m wingspan is now recognized as a fairly medial size for a Cretaceous pterosaur (the biggest species reached twice that size), a 5.5m wingspread would have impressed mid-nineteenth-century scientists who had yet to learn that extinct fliers could vastly exceed the proportions of living volant species. The fact that Bowerbank's giant pterosaur was announced in the same year that Hawkins started working on the Geological Court project is not only an indication of how cutting-edge the Crystal Palace palaeontological models were, but likely also reflects the artistic zeal that Hawkins had for the concept of reconstructing a giant pterosaur – an interest he has shared with generations of palaeoartists since.

The historic significance of *Cimoliopterus* and its presence in the Geological Court has been overshadowed by the discovery of another giant flying reptile, however – the 6–7m wingspan *Pteranodon*, discovered in 1870 in chalk deposits of Kansas (Marsh, 1871). *Pteranodon* was reported, and has been remembered, as the first giant pterosaur because its describer – the famous American palaeontologist Othniel C. Marsh – overlooked work by Bowerbank and Owen on giant pterosaur remains from Britain (Witton, 2013). Late nineteenth-century enthusiasm for American fossils and the rise of US palaeoart saw *Pteranodon* rapidly incorporated into popular texts and artwork, while Hawkins' *Cimoliopterus* sculpture remains the only major publicity that this 'original' giant species ever had. Had this reconstruction been as influential as some of Hawkins' other Crystal Palace work, the history of giant pterosaur discoveries may have been written somewhat differently.

Despite its small size and older geological age, *Pterodactylus antiquus* clearly informed the overall form of Hawkins' *Cimoliopterus*. He deviated from the *Pterodactylus* body plan in several aspects, however. While *Pterodactylus* fossils have skulls longer than their bodies, the Geological Court *Cimoliopterus* reverses this with a small head and a long, deep torso

Fig. 6.5 Details of the *Cimoliopterus* sculptures. Note the scaly skin, a feature contradicting the only direct insights into pterosaur skin known at the time of construction. (2018)

that extends in a bird-like fashion between the legs. Close inspection reveals additional bird-like anatomies, such as a localization of flight muscles on the chest (pterosaurs had flight musculature more equally shared between their chests and shoulders), avian-like legs with horizontally held thighs, and stubby, blunt tails that are poorly differentiated from the body. Although the poses of the Crystal Palace *Cimoliopterus* are not especially avian – one is posed quadrupedally, and both place the full length of their feet on the ground – these proportions make the Crystal Palace *Cimoliopterus* resemble a swan more than a traditional pterosaur.

There are perhaps two explanations for the unusual build of Hawkins' *Cimoliopterus*. The first is that Hawkins simply misread or ignored the proportional data he saw in *Pterodactylus*. If so, he was in good company: very few pterosaur artworks from the early decades of palaeontology have accurate body proportions and, even today, many artists misrepresent the real metrics of pterosaur fossils, instead reconstructing them with more intuitively pleasing bird-like forms. It seems unlikely that Hawkins was simply ignoring these fossil data, however, as his other pterosaur sculptures were proportionally closer to *Pterodactylus* (*see* below). Perhaps, instead, Hawkins was attempting to predict how the anatomy of a large pterosaur would differ from a smaller species and used large flying birds – such as swans – as a guide. Although we know today that large pterosaurs did

Chapter 6 – The sculptures: flying reptiles 101

Fig. 6.6 A twenty-first-century reconstruction of *Cimoliopterus cuvieri*, a mid-sized Cretaceous pterosaur to modern researchers, but the largest flying animal known to scientists of the mid-nineteenth century. The exaggerated arm and head proportions of this species look unusual compared to Hawkins' more bird-like sculpture, but are well evidenced and typical for large pterosaurs.

not resemble large birds, such an approach would not be without merit in the 1850s: large fliers do indeed have different muscle volumes and wing shapes to smaller species, and Hawkins would have been correct to assume that giant pterosaurs would differ in proportions to their smaller brethren.

The life appearance of pterosaurs was slowly elucidated in the decades following the opening of the Crystal Palace, but it took over a century for the true appearance of *Cimoliopterus*-like pterosaurs to become clear. Today, we recognize *Cimoliopterus* as a member of the Ornithocheiromorpha: a group of large to gigantic, long-winged, toothed pterosaurs of cosmopolitan distribution throughout most of the Cretaceous period. Although the skeletons of some ornithocheiromorphs were known in the late nineteenth and early twentieth centuries, the skulls of *Cimoliopterus*-like species were represented by fragments until Brazilian discoveries made in the 1980s revealed their full extent (e.g. Campos and Kellner, 1985). These pterosaurs had long jaws, huge, tusk-like teeth and prominent 'keel' crests at their jaw tips. Their bodies and legs were small in proportion to their heads and wings, the latter of which were long and narrow, and ideally suited to soaring over oceans (Witton, 2013). The largest ornithocheiromorphs were enormous – over 8m across the wings – but *Cimoliopterus* was probably much smaller. A rough size estimate (which is all that can be achieved, given the poor fossils of this species) is a 4m wingspan. Unbeknownst to Hawkins, therefore, *Cimoliopterus* was a more specialized and unusual pterosaur than either he or other contemporary artists were visualizing in their artwork.

'Pterodactylus bucklandi'

Owen (1854) precisely identifies the Jurassic pterosaurs as '*Pterodactylus bucklandi*', a species based on scrappy pterosaur bones from the Middle Jurassic 'Oolite' deposits of the Taynton Limestone Formation, UK (colloquially known as the 'Stonesfield Slate'). Unfortunately, little information has survived about the original Jurassic pterosaur sculptures. They were poorly documented before their disappearance almost a century ago and, other than a few dark, blurry photographs (*see* Fig. 6.2A), they are mainly represented as small objects in vintage images of the entire Geological Court. These show their general size and form, but few details can be ascertained.

Our knowledge of '*Pt. bucklandi*' has never been exemplary. Primarily represented by jaw bones, it was eventually given its own genus, '*Rhamphocephalus*', but has since been regarded as a undiagnosable, invalid species – a pterosaur represented by such poor remains that we cannot distinguish it from other flying reptiles (O'Sullivan and Martill, 2018). The poor quality fossils behind '*Rhamphocephalus*' are perhaps why Hawkins' takes on this species were so closely based on *Pterodactylus antiquus*. It is evident from photographs that Hawkins paid close attention to *Pterodactylus* proportions, resulting in large-headed, long-necked and small-bodied sculptures that are among the first proportionally accurate pterosaur artworks.

The sizes of the original sculptures are difficult to ascertain, but their 2001 replacements had heads approaching 40cm long, necks almost a metre long, and 50cm torsos, implying a wingspan of *c*. 3m. This is larger than the wingspans of the largest pterosaurs known from the Stonesfield Slate, which are estimated at 2m (O'Sullivan and Martill, 2018). Both the original and replacement models were posed standing on their hindlimbs with outstretched wings – a typical pose for early pterosaur artwork. Further details – such as skin texture, tooth shape, details of the face and so on – of the original sculptures are currently unclear.

It only took a few years for Hawkins' restoration of '*Rhamphocephalus*' to become dated. In 1859 Thomas Henry Huxley – the famous biologist best known for his defence of Darwinian evolution – realized that the Stonesfield pterosaurs better matched then-newly discovered long-tailed pterosaur skeletons from Germany. This realization implied that another pterosaur, the long-tailed, long-toothed *Rhamphorhynchus*, would have been a better model for the Geological Court Oolite pterosaurs, a view we would largely agree with today. Whatever they truly represent, the '*Rhamphocephalus bucklandi*' fossils clearly belong to the rhamphorhynchid group (O'Sullivan and Martill, 2018). It is possible that they are pieces of the Stonesfield Slate rhamphorhynchid *Klobiodon rochei*, a large (*c*. 2m wingspan), long-tailed pterosaur that probably resembled *Rhamphorhynchus* (or perhaps an earlier rhamphorhynchid, *Dorygnathus*) in overall appearance (*see* Fig. 6.7). Such animals differed from *Pterodactylus* in many respects, such as having longer, narrower wings; a long tail; a short neck; and a more robust head with strong, procumbent dentition. Were the Crystal Palace project executed today, the Oolite pterosaurs would look very different to the ones produced in the early 1850s.

Fig. 6.7 Life reconstruction of the Middle Jurassic pterosaur *Klobiodon rocheri*, a species that may have contributed in some way to the Crystal Palace Oolite pterosaurs. These long-tailed, large-toothed pterosaurs were very differently proportioned to *Pterodactylus* and would have radically changed the nature of Hawkins' sculptures, had he had access to superior fossils.

Chapter 6 – The sculptures: flying reptiles

CHAPTER 7

The sculptures: dinosaurs

For many, the main attraction of the Crystal Palace prehistoric sculptures are the dinosaurs. The four dinosaur statues of the Secondary Island are the largest models of the display and their elevation above the surrounding landscape makes them imposing, conspicuous figures even from afar. They are undeniably the most famous and spectacular components of the Crystal Palace menagerie, a status which has seen them celebrated in some quarters but ridiculed by parties out to lampoon outdated and erroneous dinosaur depictions. Such derision is entirely unfair as it ignores how much dinosaur anatomy they accurately recorded, how cutting-edge they were in the 1850s, and how progressive Waterhouse Hawkins was in his approach to their reconstruction. Though far removed from how we imagine dinosaurs now, Hawkins captured the foundation of what makes dinosaurs unique and charismatic, and even his errors reflect entirely reasonable speculation and guesswork about what was once largely unknown dinosaur form.

Three dinosaur species feature in the Geological Court: the Jurassic theropod *Megalosaurus bucklandii* and two Cretaceous taxa from the British Wealden: *'Iguanodon mantelli'* and *Hylaeosaurus armatus*. These species will be familiar to anyone interested in the early years of palaeontology as the first three dinosaurs to receive names and study from Victorian scholars, and for being the founding members of Owen's Dinosauria, which he named in 1841. Dinosaur science had taken off in earnest during the 1830s and these reptiles were already popular palaeoart subjects by the 1850s (*see* Figs 1.8–1.9), but Hawkins' creations were a major step forward for their representation in artistic and popular culture. They featured widely in promotional material and coverage of the Geological Court (*see* Figs 1.12, 3.20–3.21, 3.24) and accelerated our growing cultural fascination with all things dinosaurian.

To truly appreciate Hawkins' dinosaurs we should compare them with other early efforts to restore dinosaur form. From their first palaeoart appearance in Georg August Goldfuss' 1831 (*see* Fig. 1.12B) *Jura Formation*, the first few decades of dinosaur art featured whale-sized, lizard-like animals, with artworks like John Martin's 1837 *The country of the Iguanodon* being an especially sophisticated example of this convention (*see* Fig. 7.2). These clearly differ from the Crystal Palace dinosaurs in many ways because Hawkins' models were a significant scientific upgrade from these early works. They were more realistically sized and included many characteristic dinosaurian anatomies, including strong, upright limbs, as well as species-specific features such as beaks and armour. Hawkins incorporated a lot of mammal anatomy into his dinosaurs (*see* Fig. 2.15) and this – although not something modern palaeoartists would endorse – was also ahead of its time. In many respects dinosaurs were indeed

Fig. 7.1 The Crystal Palace *Iguanodon* and *Hylaeosaurus* as illustrated in Matthew Digby Wyatt's 1854 *Views of the Crystal Palace and Park, Sydenham*. These *Iguanodon* are surely some of the most famous pieces of palaeoart in the world – iconic not only of the Crystal Palace project, but also the early period of dinosaur research history.

Fig. 7.2 John Martin's 1837 painting *The country of the Iguanodon*, as featured in Mantell's (1839) *The Wonders of Geology*. Mantell describes this scene, where the dinosaurs are depicted as whale-sized lizards, as *Iguanodon* being attacked by a *Megalosaurus* and 'crocodile', while another *Iguanodon* and *Hylaeosaurus* confront each other in the distance. Pterosaurs, turtles, ammonites, gastropods and an ichthyosaur also feature.

Chapter 7 – The sculptures: dinosaurs **105**

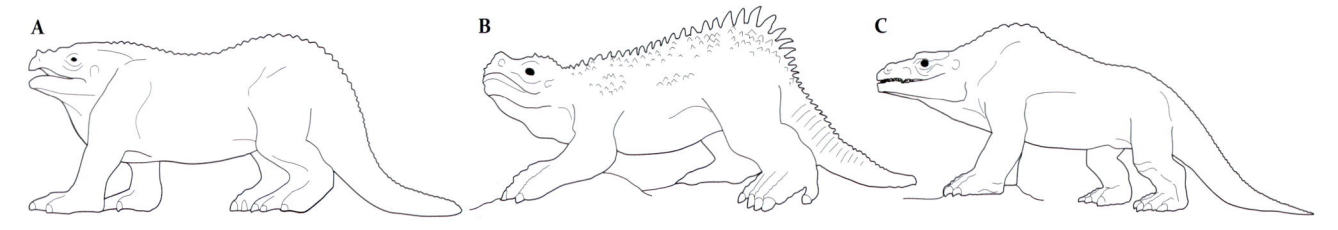

Fig. 7.3 Line drawings of the Crystal Palace dinosaur sculptures in lateral view. **A)** *Iguanodon*; **B)** *Hylaeosaurus*; **C)** *Megalosaurus*. Note how, despite sharing the same Owenian body plan, they are proportionally very different, to the extent of implying very different lifestyles and habits.

Fig. 7.4 The Crystal Palace dinosaur sculptures are sometimes ridiculed for their depiction of dinosaurs as large, heavyset quadrupeds. While inappropriate body plans for *Iguanodon* and *Megalosaurus*, such anatomy would eventually prove commonplace in dinosaurs – as shown here with the North American horned dinosaur *Centrosaurus apertus*.

more anatomically like mammals – warm-blooded, upright species – than cold-blooded sprawlers, like lizards and crocodylians. The blending of saurian and mammalian anatomy in Hawkins' dinosaurs was not lost on contemporary commentators, such as an 1854 *Hogg's Instructor* journalist:

> Instead of having short limbs and a long tapering body, like the saurians of our own time, they were curtailed in length, and increased in bulk and height, their limbs lifting the huge body high off the ground, and giving them more the appearance of the great herbivorous mammalia [sic] that succeeded them, than of any animals of their own class.
>
> ANONYMOUS, 1854, P. 284

Although a substantial haul of dinosaur bones had been amassed from southern Britain by the 1850s, no complete dinosaur skeletons had yet been found. Accordingly, the dinosaurs were constructed using Owen's concept of the dinosaur body plan: large-limbed reptiles with the features of 'heavy, pachydermal-mammals' (Owen 1842c, p. 103). This included proportionally large and erect limbs of equal length, strong, weight-bearing hips, and terrestrial habits. Each species was then augmented with features either suggested by fossil material or predicted using anatomical correlation (Owen 1842c; 1854) such that, despite their shared Owenian body plan, each genus looks remarkably different (*see* Fig. 7.3).

The results are far from the mark as goes modern interpretations of *Iguanodon*, *Megalosaurus* and *Hylaeosaurus*, but they are reasonable imaginings of dinosaurs given the material known to Waterhouse Hawkins at that time. Moreover, while the conceptualization of dinosaurs as universally robust quadrupeds was proven erroneous in relatively short order (*see* below, also Chapter 12), by the end of the nineteenth century it was clear that several dinosaur groups were large-bodied quadrupeds of some Owenian flavour: the sauropods (long-necked dinosaurs), ceratopsians (horned dinosaurs), and thyreophorans (armoured dinosaurs). The dinosaurs of Owen and Hawkins might have missed the exacting reality of these animals, but they had caught the scent of what dinosaurs were truly like and anticipated, to some extent, later dinosaur discoveries. Each model references ideas about dinosaur anatomy that were, for the time, highly modern and cutting-edge, and they thus capture the short period in history when the unique and defining aspects of dinosaurs had been recognized, but our fossil data were still too incomplete to reveal their overall forms. They represent, on a grandiose scale, an important intellectual milestone in the realization of what dinosaurs were, and are a monument to the ingenuity of early palaeontologists.

'Iguanodon mantelli'

The pair of *Iguanodon* situated atop the rise of the Secondary Island are undoubtedly some of the most famous and iconic parts of not only the Geological Court, but Victorian palaeontology in general. With total body lengths of 9.6m and shoulder heights of 3m, they are grand, impressive sculptures based on the best known dinosaur material of the 1850s. They were based on a multitude of *Iguanodon* fossils – from isolated remains to partial skeletons – that had been collected from the Wealden Supergroup: a collection of Early Cretaceous sandstones and mudrocks cropping out across the southeastern UK. The choice to create two statues may reflect the status of *Iguanodon* as the best-known dinosaur in the early 1800s as well as its prominence in early dinosaur research: Owen, Mantell and other authors had penned numerous papers on this genus in the decades preceding the 1850s. Alternatively, Hawkins may have simply felt that the composition of two giant dinosaurs looked more arresting. Whatever the reason, *Iguanodon* featured regularly in Crystal Palace promotional material in not only artwork, but also had a starring role in the 1853 New Year's Eve banquet (*see* Chapter 3).

Fig. 7.5 The famous Crystal Palace '*Iguanodon mantelli*' pair, a photograph taken by generations of visitors from the pathways around the Geological Court. (2018)

An unusually good insight into the palaeoartistry of the Crystal Palace *Iguanodon* was given by Hawkins and Owen listing the fossils that informed their reconstruction in published works (Hawkins, 1854; Owen, 1854). In doing so, they have revealed to modern readers that they were really assembling a chimera of different iguanodont remains. By the 1850s, *Iguanodon* was known from numerous sites across the Wealden Supergroup with especially extensive finds coming from the older rocks in this unit – the Hastings Beds. Hawkins and Owen treated all this material as stemming from one species, '*I. mantelli*', regardless of provenance, thus setting the foundation of a major taxonomic headache for future palaeontologists. By the end of the twentieth century, *Iguanodon* had become a notorious wastebasket of dinosaur remains. Work within the last two decades has identified at least four, but possibly up to eleven distinct iguanodonts among historic British '*Iguanodon*' material, differing in major aspects of size, proportions and detailed anatomy (Norman, 2011, 2013).

These revisions have cast the original discovery of *Iguanodon* in new light. It seems that Mantell, Owen and colleagues were actually working on multiple iguanodont taxa, including *Barilium dawsoni* and *Mantellisaurus atherfieldensis*, and not the single species '*I. mantelli*'. Indeed, fossils now recognized as genuine *Iguanodon* remains are no longer a part of the early history of British dinosaurs. Among the many issues facing late-twentieth-century *Iguanodon* taxonomists was that its type material (the name-anchoring specimen used to distinguish the species against other fossils) was anatomically uncharacteristic, and thus ill-suited to defining a fossil animal. This situation would normally see the scientific name of that animal rendered invalid and cast aside but, because *Iguanodon* is considered too important and significant a genus to abandon, its name was transferred in 2000 to a new type specimen: a completely-known, well diagnosed skeleton of the Belgian species *Iguanodon bernissartensis*. This pragmatic and sensible move saved the name *Iguanodon* from taxonomic obscurity, but also means *Iguanodon* as we define it today is no longer linked with its historic origins in the southern UK. Although *I. bernissartensis* is found in Britain, it is less common, and comes from far younger rocks than the specimens studied by Mantell, Owen and Hawkins in the early 1800s. Records show that *I. bernissartensis* fossils were actually not recorded from Britain until 1870 (Norman, 2011) and we must therefore conclude, rather unexpectedly, that no genuine *Iguanodon* fossils actually informed the Crystal Palace *Iguanodon*.

A Dentition ("*Iguanodon mantelli*" = *Barilium*, referenced by RO)

B Body size ("Great Horsham *Iguanodon*" = *Barilium*, referenced by BWH)

Expanded termination
Unexpanded termination
Curve of expanding termination

"Great Horsham *Iguanodon*" partial scapula | *Barilium* scapula | *Iguanodon* scapula

C Jaws (?*Barilium* - inferred from availability of specimens in 1852)

Dental battery
Edentulous jaw tip

D Nose horn (?*Hypselospinus* - inferred from availability of specimens in 1852)

E Proportions and general anatomy ("Maidstone specimen" = *Mantellisaurus* - referenced by RO and BWH)

i Illustrated by Mantell 1838

ii Bones represented in Maidstone specimen (based on Norman 1993)

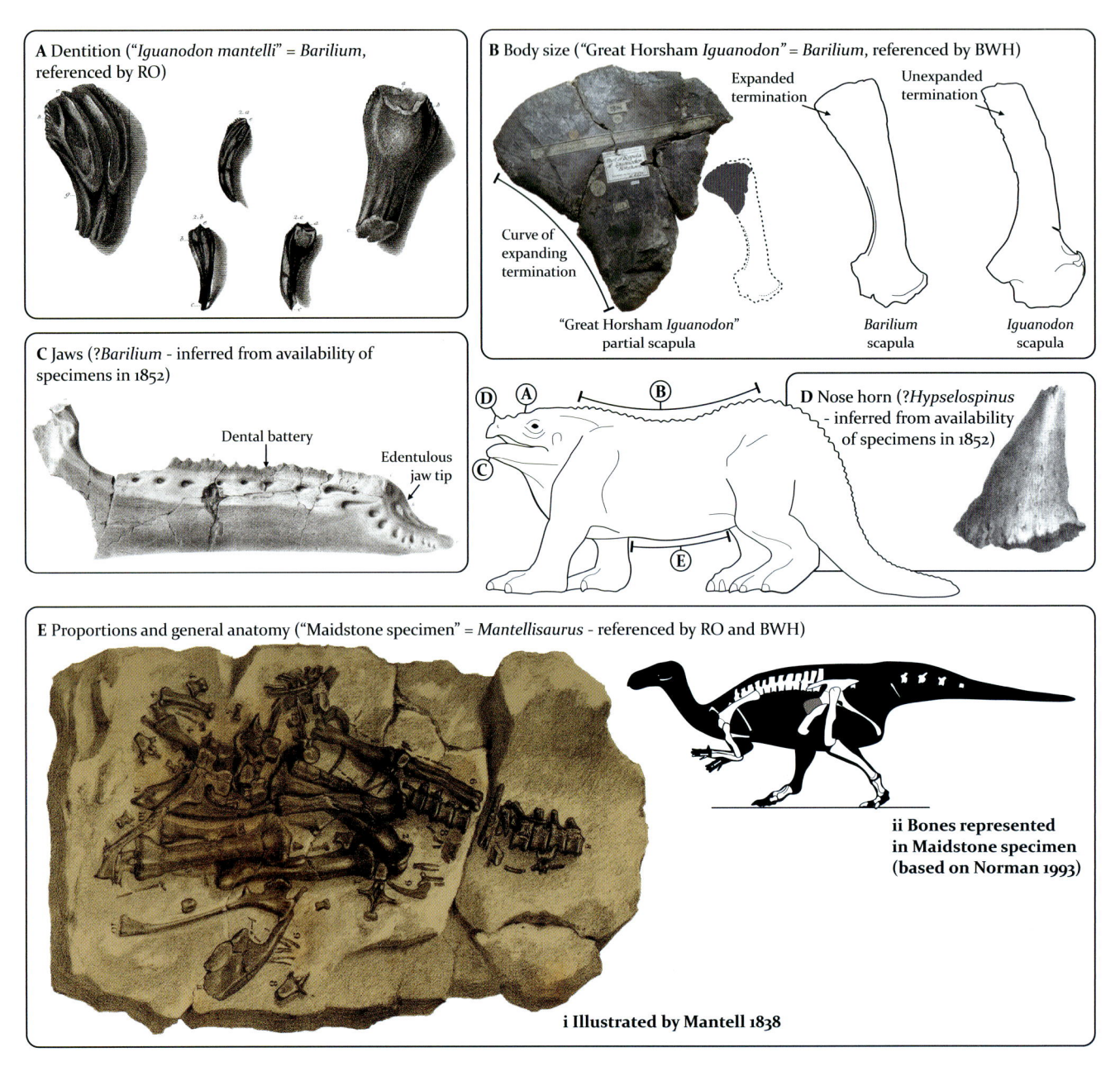

Fig. 7.6 Reference fossils likely employed in the construction of '*Iguanodon mantelli*'. *Iguanodon* teeth from Mantell (1825); jaw from Mantell (1848); 'nose horn' from Mantell (1827); 'Maidstone specimen' from Mantell (1839).

Recent characterization of British iguanodont anatomy allows us to identify which species were actually factored into the Crystal Palace designs. The fossils mentioned by Owen (1854) and Hawkins (1854) are probably not a comprehensive list of consulted materials but allude to at least three important reference specimens. Firstly, Owen (1854) identified the models as being '*I. mantelli*', one of the two names (the other being '*I. anglicus*') applied to the original set of *Iguanodon* teeth described by Mantell in 1825. These teeth are a close match for those from the large iguanodont we now call *Barilium dawsoni* (Norman, 2011). Hawkins (1854) mentioned that the size of the models was based on a robust specimen with the nickname of the 'Great Horsham *Iguanodon*': a collection of iguanodont limb, rib and shoulder material from Horsham, Surrey. This largely forgotten specimen was instrumental in Owen's downgrading of *Iguanodon* size from a whale-like behemoth to a mere 10m long (Owen, 1855) and is almost certainly another example of *Barilium*, due to it being sourced from the Hastings Beds and bearing several characteristic features of this taxon (*see* Fig. 7.6B;

see Norman, 2011, 2013 for discussion of *Barilium* anatomy and distribution).

Both Hawkins (1854) and Owen (1854) reference a third specimen, a disarticulated partial iguanodont skeleton from Maidstone, Kent. This specimen does not seem linked to any one aspect of the sculptures and may have acted as a general reference for proportions and form. Today, the specimen is tentatively regarded as an example of *Mantellisaurus atherfieldensis* (Norman, 1993, 2011). This fossil was studied extensively by Mantell (1839, 1851), as well as by Owen (1842c, 1851) and is known colloquially as the 'Mantel-piece' or 'Maidstone specimen'. Found in 1834 and representing the first discovery of substantial dinosaur remains, the Maidstone specimen provided important early insights into dinosaur body plans and was the basis for Mantell's earliest reconstructions of *Iguanodon* (Norman, 1993).

It's fair to say, however, that potentially significant aspects of the Maidstone fossil were overlooked by Mantell, Owen and Hawkins. Iguanodont expert David Norman (1993) found that this jumble of bones actually constituted much of the body, including the majority of the torso vertebrae, limb girdles, limbs, ribs and some tail elements. Critically, it clearly showed that iguanodont forelimbs were shorter and more gracile than those of the hindlimbs (*see* Fig. 7.6E). This fact seems to have completely bypassed Owen, who remarked in 1842(c) that the *Iguanodon* humerus had yet to be 'satisfactorily determined' (p. 135), but not Mantell:

> …the hinder extremities, in all probability, resembled the unwieldy contour of those of the Hippopotamus or Rhinoceros, and were supported by strong, short feet, protected by broad ungual phalanges: the fore feet appear to have been less bulky, and adapted for seizing and pulling down the foliage and branches of trees…
> MANTELL, 1851, P. 313

Although implying that *Iguanodon* forelimbs had a role beyond support (as seems to be the principal implication in the Crystal Palace models), Mantell's observations stopped some way short of declaring *Iguanodon* a bipedal animal, as later discoveries would prove. Had the Maidstone specimen been examined more diligently by Hawkins and Owen, the form of *Iguanodon*, and perhaps other dinosaurs, might have been apparent much earlier, and the Crystal Palace models would have looked very different.

In addition to these specimens listed by Hawkins and Owen, at least two other fossils were probably factored into the Crystal Palace sculptures. Firstly, the beaked, toothless jaw tips were likely taken from a jaw specimen described by Mantell in 1848. *Iguanodon* jaw bones were almost entirely unknown before the 1850s such that Hawkins almost certainly referenced the significant and timely discovery of this first near-complete lower jaw. This specimen has been given various names over the years, but is now generally considered to represent *Barilium* (Norman, 2013).

Secondly, it is not clear exactly which specimens inspired the nose horns (the famously misinterpreted thumb spike) as several were known by the 1850s, but the horns of the sculptures and associated artworks are somewhat recurved, making them similar to a large, recurved 'nasal horn' published by Mantell in 1827. Iguanodont thumb claws vary between species in details such as size and curvature, and from these we can rule out *Iguanodon*, *Mantellisaurus* and *Barilium* as likely anatomical references. Rather, the large and recurved 'nose horn' morphology best fits the thumb spike of yet another Wealden iguanodont species, *Hypselospinus fittoni* (*see* Norman, 2015 for details of this animal).

We thus have evidence of at least two, and possibly three, iguanodont species informing the Crystal Palace *Iguanodon* and, with multiple taxa contributing to their form, no single generic label can be satisfactorily given to the sculptures today. They are clearly best viewed as chimeras of multiple iguanodont species. If forced to select a single most appropriate generic title however, *Barilium* may be our best bet (*see* Fig. 7.7). It was not only the anatomical reference for the teeth, jaws and overall size, but was also a very large, robust, and perhaps even wholly quadrupedal (Norman, 2011) iguanodont stemming from the Hastings Beds deposits – the same rocks which were so crucial to early work on *Iguanodon*. *Barilium* thus embodies many characteristics of the Crystal Palace *Iguanodon*, and might be the closest real-life species to the animal Hawkins was aiming to restore.

The near 10m body lengths of the *Iguanodon* models were, for the time, very modern interpretations of dinosaur size. Earlier length estimates based on scant bones extrapolated *Iguanodon* to many tens of metres long (*see* Fig. 7.2; Mantell, 1851; Owen, 1854, 1855). Despite factoring Owen's downsizing, the *Iguanodon* are among the biggest of denizens of the Geological Court and are also some of the most structurally complex. Moulded from 30-tonne clay sculptures, the final models were composed of brick, tile, cement and iron. Hollow cores crossed with iron struts make the models lightweight and economical for their size, while metal teeth, cast to resemble the first *Iguanodon* teeth described by Mantell, are socketed into their jaws. Hawkins (1854) used the standing *Iguanodon* as an example of the mechanical

Fig. 7.7 With the Geological Court *Iguanodon* actually being a chimera of several species, no one taxon can be chosen as their 'true' identity. *Barilium dawsoni*, a large iguanodont from the southern UK, is probably the best representative of the sort of animal Hawkins and Owen were trying to reconstruct, however. Fossils of *Barilium* informed much of their reconstruction and its large and robust physique matches the physical presence of the Geological Court models.

challenges involved in constructing such heavy models on legs ('not less than building a house upon four columns', p. 447) without the aid of disguised supporting agents. The forward-reaching right arm of the resting sculpture appears to have defied this intention and uses an artificial cycad trunk as a rest, however.

Hirst Conservation's 2000 paint analysis suggests that green or green/blue paint was originally applied to the *Iguanodon*, with historic photos suggesting a relatively uniform patterning. Towards the end of the twentieth century both sculptures were painted dark green with sharply contrasting white undersides and eye regions but, since 2001, they have sported relatively muted greens.

Although both Mantell and Owen wrote extensively about *Iguanodon* before 1854, their comments on its likely life appearance were relatively brief and Hawkins had considerable room for anatomical interpretation. He seems to have assumed that their herbivorous habits justified especially large, expansive guts, such that their torsos are deeper and wider than that of the carnivorous *Megalosaurus*. Their limb proportions are elephantine, in that the upper portion of the limb is the longest element, and they are not as thickly muscled as those of the *Megalosaurus*: Hawkins seems to have imagined *Iguanodon* anatomy as being better suited to supporting a vast body than running fast. The asymmetrical shoulders of the standing sculpture indicate mammal-like mobile shoulder blades and a low hump over the shoulders suggests Hawkins was imagining something like the expanded neck musculature of large-headed mammals, such as certain bovids and rhinoceros, to support the heavy head and jaws. Both *Iguanodon* are well-ornamented: they have large 'feature' scales along their spines and in a short row behind the head, a small dewlap, as well as a pair of nose horns. The latter is not a reporting mistake: although the models are often described as having singular nose horns, both sculptures have smaller secondary horns sitting behind the more obvious examples at the jaw tip.

The nasal horns of the *Iguanodon* are often mentioned as examples of the inaccuracy of the Geological Court palaeontological sculptures, but they were actually a controversial addition even in 1854. Owen was strongly opposed to *Iguanodon* having such a structure and described it as 'more than doubtful' in his 1854 guidebook. A year later, Owen devoted several pages of his *Iguanodon* monograph to dismantling the nose horn identification and correctly identified the 'horn' elements as spike-like claws (Owen, 1855). Hawkins was thus following Mantell's (1833, 1839) interpretation of *Iguanodon* appearance, which itself was based on the presence of small nasal horns in several living iguana species.

Iguanas may have informed the models in other regards as well, including the large tympanum (the conspicuous 'ears'), dewlaps, the midline feature scales, and the large, non-overlapping scales of the face. The edentulous beak and blunt claws are deviations from iguanas and presumably

Fig. 7.8 Details of the Crystal Palace *Iguanodon*. Maintenance holes in the bottom of the standing sculpture allow for access to a now heavily reinforced internal cavity. (2018)

reflect Hawkins following *Iguanodon* specimens: on the former, Hawkins was more accurate than Mantell, whose final conceptualization of *Iguanodon* included prehensile lips and a protruding tongue (Mantell, 1851). The grasping hand of the reposed *Iguanodon* could be a Mantellian nod, however, reflecting his proposal that *Iguanodon* used its forelimbs to grasp vegetation (Mantell, 1849, 1851; Dawson, 2016). If so, it's noteworthy that Hawkins did not follow Mantell's observation that the forelimbs were relatively slender, instead sticking to the Owenian idea of *Iguanodon* being a columnar-limbed quadruped. The grasping hand may merely be Hawkins making one *Iguanodon* a little more animated, just as he showed groups of animals in various poses in other parts of the Geological Court.

Whatever the reason, the posing of the *Iguanodon* counts them among the most lifelike sculptures in the Geological Court. Most of the other reptiles are posed fairly functionally, lying or standing in ways that allow us to see their form clearly. But the *Iguanodon* seem frozen in a moment of time: the reposed individual is relaxed with squatting hindquarters while pawing at a cycad, and the standing animal, positioned above its partner, appears vigilant and alert. Their arrangement implies real behaviours and personalities in these anatomically fictional restorations, and the difference in displayed soft tissues – specifically, the configuration of their bellies and musculature – enhances their portrayal as real animals. The belly of the reclining *Iguanodon* is spreading under its weight, while the musculature of the hindlimbs is visibly deformed by the crouching pose (*see* Fig. 2.15). Whether visitors notice this consciously or not, these subtle differences imbue a sense of believability and personality to these restorations.

Just over a decade after the Crystal Palace dinosaurs were installed, discoveries of another herbivorous dinosaur in New Jersey, USA began to erode the credibility of Hawkins' *Iguanodon*. This new animal, *Hadrosaurus*, was evidently bipedal due its very short forelimbs, and thus looked very different to what was being imagined in Victorian Britain (of note is that Hawkins was directly involved in reconstructing *Hadrosaurus* in the United States during the 1860s – an act that simultaneously undermined his Crystal Palace work while also advancing his reputation in the restoration of

Chapter 7 – The sculptures: dinosaurs

fossil animals; Bramwell and Peck, 2008). Later, a suite of complete *Iguanodon* skeletons discovered in Belgium during 1878 showed how wide of the mark Hawkins' *Iguanodon* were. These skeletons, recovered from a Bernissart coal mine, showed an animal with long, powerful hindlimbs; relatively slender, short forelimbs equipped with large thumb spikes; a beaked, horse-like head with sophisticated dental batteries for chewing plant matter; and a long, deep tail. Initially restored as a kangaroo-like biped, *Iguanodon* would eventually be realized as we know it today: a horizontally-backed ornithopod dinosaur that could alternate between bipedal and quadrupedal gaits.

The ornithopods, a widely dispersed and very successful group of herbivores, are now some of the most studied and best-known of all dinosaurs thanks to their exceptional fossil record, which includes whole skeletons preserved sheathed in skin. Although complications in the history of *Iguanodon* taxonomy muddy exactly what iguanodont species was restored at Crystal Palace, it remains true that *Iguanodon* is the only Geological Court dinosaur now known from complete skeletons. Comparing these and other ornithopods with their 1854 sculpted counterparts is a poignant indicator of how far palaeontological science has moved since the era of Hawkins and Owen.

Hylaeosaurus armatus

The most overlooked dinosaur of the Crystal Palace trio is *Hylaeosaurus armatus*: the first ankylosaur (a type of armoured dinosaur) known to science. The fact that *Hylaeosaurus* is very much the third, somewhat forgotten dinosaur species of the Court is unfortunate given that it was Hawkins' most insightful dinosaur reconstruction (Norman, 2000, 2011). Like *Iguanodon*, *Hylaeosaurus* was a Wealden species worked on primarily by Gideon Mantell and its remains were found in the same Sussex queries that yielded the first iguanodont fossils. The original *Hylaeosaurus* specimen, discovered in 1833, was a noteworthy find for being an articulated portion of skeleton – the first specimen of a dinosaur represented by more than isolated bones or disarticulated remains. Alas, these remains were uncovered by quarry blasting, a destructive process which tends to destroy parts of fossils in the process of exposing them. At least a third of the specimen is thought to have been lost in the explosion that also revealed it (Blows, 2015) and a similar fate befell other *Hylaeosaurus* remains discovered in the early 1800s. Had they been uncovered more delicately, there's a chance *Hylaeosaurus* would have been more informative to Victorian science than *Iguanodon*.

Like the other dinosaur sculptures, the *Hylaeosaurus* is constructed of brick, tile and cement, and has a hollow core spanned by iron struts. At 8.4m long, it is the smallest of the four dinosaur models. During the mid-1970s or earlier, the head fell off the sculpture and a significantly lighter fibreglass replica was installed as a replacement in 1976. The original head was moved to the park visitor centre (McCarthy and Gilbert, 1994) before being relocated again to the vista overlooking the Geological Court (Fig. 7.9C). For visitors arriving from Crystal Palace train station or Anerley Hill, this curious, disembodied face is the first sculpture encountered on approach to the Geological Court and immediately draws guest attention, but is rarely recognized as a historic artefact. Concerningly, unlike the other palaeontological sculptures it lacks any barrier or protection from the public so it is routinely climbed over and sat on by park visitors while the surrounding area is eroded and turns to a mud- or dust-bath for much of the year. Though made of robust materials, there are obvious conservation concerns around this irreplaceable and significant piece of vintage art being used as a bench and climbing frame.

The first *Hylaeosaurus* remains were initially thought to be a portion of *Iguanodon* and it was only the presence of osteoderms (a bony plate or spike embedded in skin, often for defensive purposes) which convinced Mantell that he was dealing with an entirely new kind of dinosaur. This amounted to the third dinosaur species known to science, and it was named and described in 1833 as *Hylaeosaurus armatus*. But unlike *Iguanodon* or *Megalosaurus*, *Hylaeosaurus* never became an especially popular fossil reptile and, even in the Geological Court, it's easy to ignore due to the sculpture facing away from the paths around the Secondary Island. The showcasing of the hindquarters affords excellent views of the armoured back, but the obstructed view of its face undeniably leaves a lesser impression on visitors, especially as it competes for attention with the larger, fully visible dinosaur statues either side. There are accordingly fewer visual records of *Hylaeosaurus* from the last 170 years and details of its colouration and patterning over this time are scarce. In 2000 Hirst Conservation found that the original colour was probably a rich green with a brown glaze. By the 1990s, this had transformed into a dark green with a white belly and, since 2001, it has adopted a more muted, leaf green.

Although ankylosaurs are now very well researched dinosaurs, *Hylaeosaurus* remains an enigmatic animal. It is thought to be a nodosaurid, a group of Jurassic and Cretaceous

Fig. 7.9 *Hylaeosaurus armatus*, the Crystal Palace ankylosaur in 2018. **A)** as seen from the pathways around the island without a visible face; **B)** as seen from the front, revealing the extremely detailed face mostly unseen to generations of park visitors. This head is a fibreglass replacement of the original; **C)** the original, concrete *Hylaeosaurus* face situated close to the Geological Court, 2013.

ankylosaurs which lacked a tail club, and was probably closely related to the better known Wealden nodosaurid *Polacanthus*. Like other early nodosaurids, it probably bore several types of armour, including a distinctive shield over its hip region. A number of fossils have been referred to *Hylaeosaurus* in the last two centuries but modern palaeontologists struggle to distinguish them from other armoured dinosaurs, such that the only specimen incontrovertibly regarded as representing this species is the original slab described by Mantell in 1833 (Barrett and Maidment, 2011; but *see* Pereda-Suberbiola, 1993 and Blows, 2015 for differing opinions). This specimen, exposed on its underside, contains the posterior portion of the skull, several neck and shoulder vertebrae, elements of the shoulder girdle and the large spike-like scutes that covered its skin (Barrett and Maidment, 2011). In the mid-1800s *Hylaeosaurus* inventories were much broader, however, and included referred jaws, teeth, limb bones, hip vertebrae and an articulated series of armoured tail vertebrae (Owen, 1842c). This rendered *Hylaeosaurus* c. 1854 comparable to other dinosaurs in fossil representation, and Hawkins was able to capture many distinguishing features of armoured dinosaurs in his sculpture.

It is probable that at least some of the material used by Hawkins was not true *Hylaeosaurus*. Many specimens once referred to *Hylaeosaurus* are now regarded as indeterminate remains of nodosaurids (Barrett and Maidment, 2011) and some were from very different types of dinosaur: specifically, the jaws and teeth Owen referred to *Hylaeosaurus* are from a stegosaur and sauropod, respectively (Upchurch *et al.*, 2011; Barrett and Maidment, 2011). The idea that *Hylaeosaurus* inventories were mixing dinosaur types was actually realized by Mantell in 1848, who erected a new genus for the jaw specimen: '*Regnosaurus northamptoni*' (a species no longer considered valid, on account of its very poor remains). Owen was unmoved by Mantell's arguments, however, and retained this material in *Hylaeosaurus* (Owen, 1858). Details of the sculpture suggest that Hawkins followed Owen into his *Hylaeosaurus* reconstruction, making it a chimera of stegosaur and nodosaur body parts.

The most obvious and striking aspect of the *Hylaeosaurus* sculpture is its armour. Hawkins' arrangement of this in a single row was reasonable given the material he was working with, but was nevertheless controversial even in 1854. Two types of osteoderm were recognized for *Hylaeosaurus* in the 1830s and '40s: large spikes, and low elliptical structures that sometimes bore tubercles at their summit. Both feature prominently on the upper surface of Hawkins' sculpture, covering the braincase, shoulders, back, haunches and tail. As with modern armoured and spiked creatures, Hawkins regionalized the armour in ways to protect more exposed parts of the body, such as the shoulders, flanks and hip region. Ankylosaur researcher William Blows (2015) commended Hawkins' attention to detailing the facial armour of *Hylaeosaurus*, noting that the distribution of scales and prominences was reminiscent of the scalation patterns and developmental emphasis of real ankylosaur skulls.

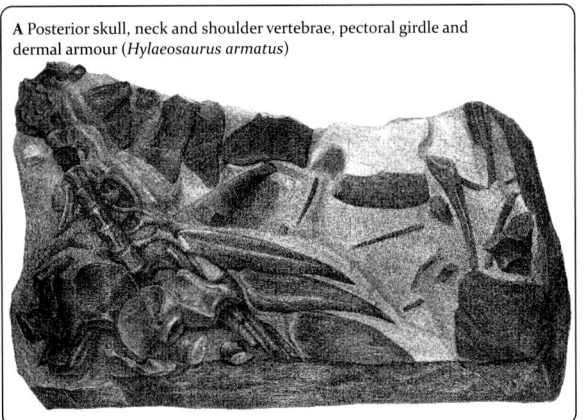

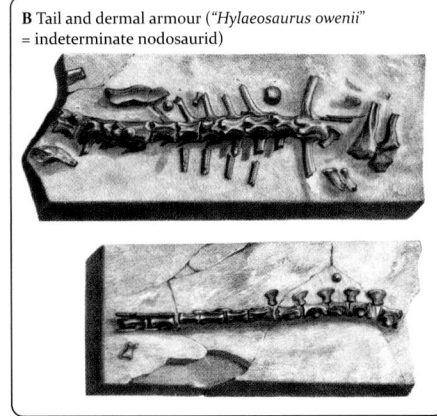

Fig. 7.10 Likely reference fossils for the Crystal Palace *Hylaeosaurus armatus*. Much of the material here is not *Hylaeosaurus*, but was considered to belong to this genus in the 1850s. Images of fossil specimens from Owen (1884).

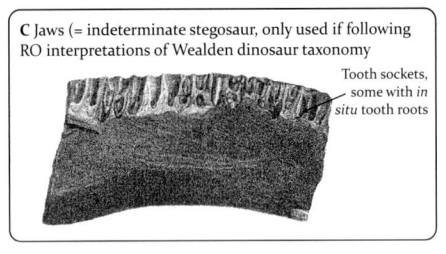

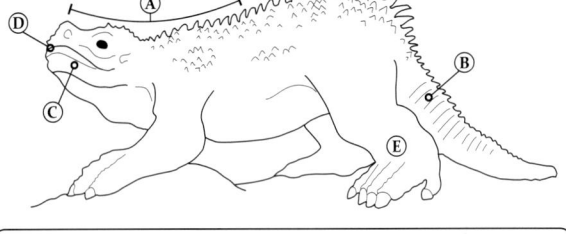

A Posterior skull, neck and shoulder vertebrae, pectoral girdle and dermal armour (*Hylaeosaurus armatus*)

B Tail and dermal armour ("*Hylaeosaurus owenii*") = indeterminate nodosaurid)

C Jaws (= indeterminate stegosaur, only used if following RO interpretations of Wealden dinosaur taxonomy
Tooth sockets, some with *in situ* tooth roots

D Teeth (= sauropod teeth, referred to *Hylaeosaurus* by RO but not featured in sculpture)

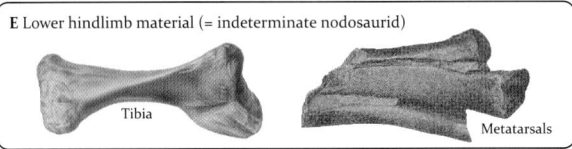

E Lower hindlimb material (= indeterminate nodosaurid)
Tibia
Metatarsals

The single row of spines used by Hawkins was conceived by Mantell (1833), who drew comparisons between the armour of *Hylaeosaurus* and the spiny midline fringe of iguanine lizards. Owen (1842), however, held reservations about the osteoderms being armour at all. While not fully disagreeing with Mantell's interpretation, Owen proposed that the 'spines' could also be displaced abdominal ribs (unlike mammals, reptiles possess ribs lining their belly regions, embedded in the soft tissues). The Geological Court guidebook shows that, by 1854, Owen had conceded that Mantell's armour hypothesis was more plausible, but he also remarked that Hawkins' depicted spines configuration was conjectural. When Owen returned to this issue again in 1858 he provided sound reasoning for the spines being arranged in two parallel rows along the back, and in doing so predicted something close to the reality of many ankylosaurs. What both Owen and Mantell missed, however, was that the spines in the original *Hylaeosaurus* specimen had not moved far from their in-life position: complete nodosaurid specimens have since shown that these animals were covered in multiple rows of scutes, spines and spikes, with the largest spines projecting from the sides of their necks, bodies and tails (*see* Fig. 7.12).

The Crystal Palace *Hylaeosaurus* is the most conventionally reptilian of the three featured dinosaur species. Although its limbs are positioned in an upright fashion, it has a low, crouched pose, a wide body and a short, broad face. The hands and feet are large with ankles and wrists placed entirely on the ground, and the feet possess an entirely non-dinosaurian long, splayed fifth toe. The face is lizard-like in having large scales in proportion to the body, including obvious lip scales. The rugose, tough-looking scales covering the face and body are reminiscent of the skin of heavyset iguanines, such as the Galapagos genera *Amblyrhynchus* or *Conolophus*, particularly with the presence of low tubercles and hornlets around the back of the head, the scaly nasal prominence, and polygonal scales on the body and limbs. This may reflect Hawkins following Mantell's (1833) referencing of iguanas as a model for *Hylaeosaurus* skin. Despite the presence of a beaked jaw tip in *Iguanodon*, Hawkins restored an entirely lipped, lizard-like mouth for *Hylaeosaurus*: complete ankylosaur jaws would eventually reveal that they too had beaked jaws. The distinctive line of the mouth – which downturns towards the jaw tip – is not especially lizard-like, however, and might reflect Hawkins factoring the bowed '*Regnosaurus*' jaw specimen into the sculpture. If so, Hawkins may have been choosing interpretations from different scientists for the same animal, following Owen for the jawline, but Mantell for the armour arrangement.

Fig. 7.11 Additional details of the *Hylaeosaurus* sculpture, from 2018. Note the presence of a nasal boss and enlarged, thickened scales over the neurocranium, recalling certain iguana species. The fibreglass material of the head is obvious from the internal view.

Fig. 7.12 A modern reconstruction of *Hylaeosaurus armatus*: a wide, low-bodied, beaked species covered in rows of spines and a shield-like structure over the hips – in many respects, not too dissimilar to the Crystal Palace reconstruction. Much of this restoration is based on the closely related *Polacanthus*, as *Hylaeosaurus* remains a poorly represented species.

Hylaeosaurus is Hawkins' only dinosaur sculpture without visible dentition. Given that Owen (1842, 1854) had referred teeth to this species, and that virtually all the other non-mammal statues have exposed dentition, this omission is peculiar. Perhaps, if the sculpture was always planned to face away from visitors, sculpting an open mouth with socketed teeth may have been unnecessary. Moreover, the teeth thought to belong to *Hylaeosaurus* were small and fine: perhaps they were too delicate to cast or set into the jaw in the way achieved for the other dinosaur sculptures. Of course, more basic factors – time, money, artistic whim – may explain this too.

Hawkins' *Hylaeosaurus* is his most authentic dinosaur with respect to our modern understanding, accurately

Fig. 7.13 The first near-completely known Mesozoic dinosaur, *Scelidosaurus harrisonii*. Discovered in 1858 and described by Owen in 1863, this quadrupedal, armoured species vindicated Owen's concept of ankylosaur life appearance as well as (temporarily, at least) the dinosaur body plans seen at Crystal Palace. Neither Hawkins nor Owen seem to have capitalized on this discovery to promote their Geological Court work, however.

capturing an ankylosaur as a low-slung quadruped with a short neck, long tail and an armoured back. Unlike the Crystal Palace *Megalosaurus* and *Iguanodon*, which were soon embarrassed by the discovery of fossils indicating entirely different body forms, the depiction of *Hylaeosaurus* as a heavyset quadruped was vindicated in 1858 when *Scelidosaurus harrisonii* was found in Jurassic deposits of the southern UK (*see* Fig. 7.13). This specimen – the first essentially complete dinosaur skeleton ever found – matched many aspects of the Crystal Palace *Hylaeosaurus* as well as Owen's general ideas about dinosaur form.

Curiously, despite describing *Scelidosaurus* in detail, Owen (1863) did not capitalize on this opportunity for personal vindication and promotion of the Crystal Palace Dinosaurs. Norman (2000) ascribes this to Owen being ever busier from the late 1850s onwards, taking on prominent roles in establishing the British Museum (Natural History) (now the Natural History Museum, London), participating in numerous debates about evolution and natural selection, as well as maintaining a high output of papers, books and monographs. Owen's general disinterest in the Crystal Palace (*see* Chapter 2) surely has relevance here as well, as might the limited success of the wider Crystal Palace Park project (*see* Chapter 12). Hawkins' lack of response to *Scelidosaurus* is also peculiar, as he also never seems to have championed it as an example of his anatomical insight as he did with some other dinosaur discoveries (*see* below). By the 1860s Hawkins was busy working on other dinosaur material in the United States, however, and dinosaurs in general were becoming better known. Perhaps these projects eclipsed Hawkins' view of new dinosaur discoveries in the UK, even those which vindicated his anatomical predictions.

Megalosaurus bucklandii

The only carnivorous dinosaur in the Geological Court is *Megalosaurus*, an imposing, muscular and massive (*c.* 10.7m-long) sculpture that occupies the central portion of the Secondary Island. It is the oldest dinosaur represented in the site in more ways than one, representing both the first dinosaur to receive scientific study (being named by the famous geologist Rev. William Buckland in 1824) as well as the geologically oldest species. In stemming from Middle Jurassic rocks of Oxfordshire, *Megalosaurus* was about 40 million years older than the other dinosaurs at the site. But while not thought to be a contemporary with *Hylaeosaurus* or *Iguanodon* in 1854, the arrangement of the sculptures is such that it's difficult not to imagine this large predator eyeing the adjacent dinosaurs as a potential meal. The discovery of dinosaur carnivores erroneously referred to the genus *Megalosaurus* in Wealden rocks closed this time gap and seemingly justified the glut of nineteenth-century palaeoart showing *Megalosaurus* battling *Iguanodon*. After Crystal Palace Park opened, many of these scenes incorporated designs clearly based on Geological Court artwork (*see* Chapter 12).

The Crystal Palace *Megalosaurus* is one of the most iconic parts of the Geological Court and it is easily among the most oft-depicted components of the display. As with all the sculptures, full details of its historic colouration aren't clear.

116   Chapter 7 – The sculptures: dinosaurs

Fig. 7.14 The Crystal Palace *Megalosaurus bucklandii*, as seen in 2013.

A Lower jaw and dentition (*Megalosaurus bucklandii*)

B Metacarpal (= *Megalosaurus bucklandii* metatarsal)

C Pelvic element (ilium) and sacral vertebrae (*Megalosaurus bucklandii*)

E Dorsal verteba and ribs (*Megalosaurus bucklandii*)

D Caudal vertebra (*Megalosaurus bucklandii*)

F Femur (*Megalosaurus bucklandii*)

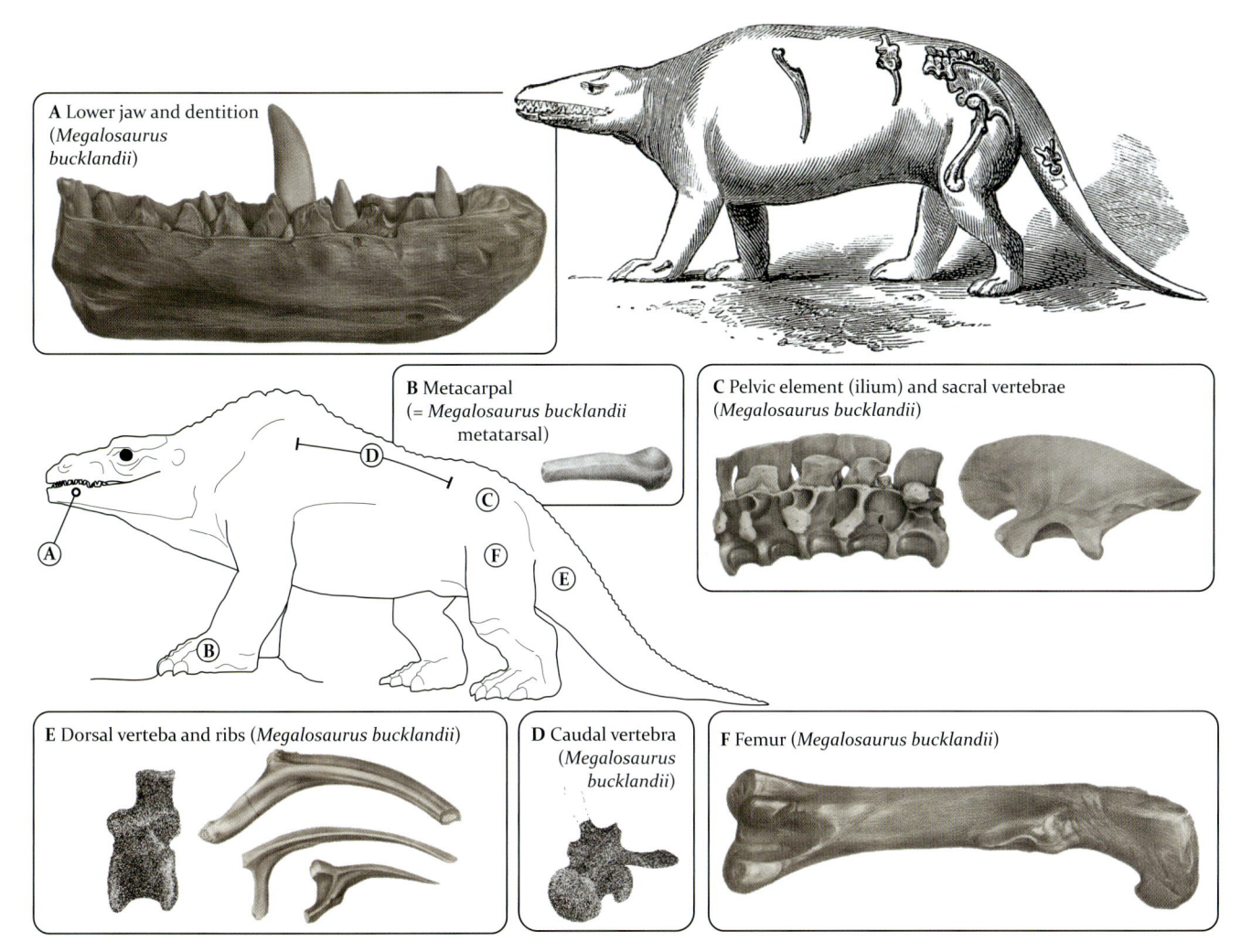

Fig. 7.15 Fossil material incorporated into the Crystal Palace *Megalosaurus bucklandii*. Owen's (1854) skeletal and life reconstruction features in the top right: although sharing a similar body plan to Hawkins' sculpture, it possesses several important anatomical differences. Fossil images from Buckland (1824) and Owen (1884).

Fig. 7.16 Details of the Crystal Palace *Megalosaurus*, as seen in 2018. Note the inset metal teeth, concealed abdominal hole for maintenance access and drainage, and interior brickwork construction.

Hirst Conservation (2000) concluded that it was originally painted with the same terre-verte green and brown glaze as the *Hylaeosaurus*, but later archive photographs and video footage mainly show lighter, unpatterned colour schemes, sometimes with subtle face striping and countershading. It was by far the most ostentatious sculpture in the Geological Court during the 1970s, however, when it was painted a garish reddish-brown – a somewhat bizarre restorative choice given the otherwise relatively muted tones characterizing the sculptures. Since at least the early 1990s the Crystal Palace *Megalosaurus* has been various shades of grey-brown.

Hawkins' *Megalosaurus* is specifically meant to represent *M. bucklandii*, a species known to him from a jaw bone and teeth, a few vertebrae and ribs, pelvic bones and some hindlimb material (*see* Fig. 7.15). It is a fairly faithful representation of Owen's vision of *Megalosaurus*, which, uniquely among the dinosaurs, was illustrated in his 1854 guidebook (*see* Fig. 7.15A). This famous image is also one of the oldest known skeletal diagrams of a dinosaur and does not sugarcoat the volume of material then known for *Megalosaurus*. It stands in marked contrast to Owen's attempt at championing the scientific authenticity of the Crystal Palace reconstructions, which he stated were based on species which 'the entire, or nearly entire, skeleton had been exhumed in fossil state' (Owen, 1854, p. 5). Curiously, Owen did not reference the full 1854 inventory of *Megalosaurus* fossils in his drawing or text, omitting some potentially significant vertebrae from rocks of the Wealden Supergroup that had been referred to this taxon. This suggests that Hawkins' *Megalosaurus* was based entirely on Jurassic material that we still regard as *M. bucklandii* today, and that it narrowly escaped being a chimera like the other Geological Court dinosaurs. By the 1850s, the *Megalosaurus* name was being applied to just about any new carnivorous dinosaur fossil, and it soon became another Victorian wastebasket taxon. Over time, the *Megalosaurus* name grew to contain dozens of species of varying validity, and it was only in 2010 that this was sorted through. Detailed study of this bloated taxonomic inventory found a number of previously unrecognized carnivorous dinosaur species, but only *M. bucklandii* was maintained as a valid *Megalosaurus* species (Benson, 2010).

Hawkins' *Megalosaurus* has several hallmarks of predatory species that reflect close observation of living animals. The limbs, for instance, are not pillar-like as in his *Iguanodon*, but have flexed joints and bulging musculature. They recall the limbs of rhinos, which are relatively fast, sprightly animals despite their large size, more than the columnar limbs of slower-paced giants, like elephants. Although the proportions of the hindlimb are somewhat elephantine, in that the thigh region is relatively long, the forelimb has a short upper region which, in concert with the short, sharply clawed hands and feet, recalls the arms of bears. The body is considerably more trim and streamlined than those of the other dinosaurs, tapering from the muscular shoulders towards the hips, and it also lacks an expansive gut. Hawkins would have known that carnivorous animals have smaller, shorter guts than herbivores, and he probably modelled his *Megalosaurus* with this in mind. The massive head (some 1.5m in length, probably based on Owen's older calculations from 1842, and not reflecting the revised 2'6" (0.76m) length published by Owen in 1856) is held in place with deep neck muscles anchoring to an enlarged shoulder skeleton, likely in reference to large-headed mammals such as bovids and rhinos.

Today, it's clear that Owen and Hawkins overestimated the size of the *Megalosaurus* head, but their prediction that large-headed carnivorous dinosaurs needed augmented head support has been borne out in several species (see, for

Fig. 7.17 The *Megalosaurus* shoulder hump, a speculative structure added by Hawkins that was seemingly (though not actually) vindicated by fossils potentially known to Owen as Hawkins worked. **A)** the high-spined *Altispinax dunkeri* vertebrae that Owen (1855, 1856) regarded as representing the shoulder region of *Megalosaurus*; **B)** the shoulder hump of *Megalosaurus* in detail (note the bulging musculature); **C)** the likely reality of the *Altispinax vertebrae*: they actually belong to the back of the torso, not the shoulders, and were likely part of a sail-like structure. The theropod in this scene, *Concavenator corcovatus*, is a completely known predatory dinosaur with a posterior torso sail, and *Altispinax* may have resembled it. A, from Owen (1884).

example, the expanded anatomy for receipt of large neck muscles on the skull of *Tyrannosaurus*: Snively and Russel, 2007). The collective impression of Hawkins' *Megalosaurus* is a creature that looks undeniably powerful and predatory: a mix of bear, buffalo and crocodile.

Having mentioned the shoulder hump (*see* Fig. 7.17), it would be remiss not to outline the interesting history of this structure, as it sheds light on the dynamic between Owen and Hawkins. Naish (2010) and Lomax and Tamura (2014) assumed that the shoulder hump of *Megalosaurus* was based on the poorly known remains of *Altispinax dunkeri*, a predatory species represented by three tall-spined Wealden vertebrae known to Owen in the 1850s. Owen (1855, 1856) regarded these as the shoulder vertebrae of *M. bucklandii* and remarked that their tall spines must have anchored powerful, head-supporting muscles which allowed *Megalosaurus* to pull apart carcasses via strong movements of its head and jaws. It's logical to assume that Owen gave Hawkins advance notice of these new data but, surprisingly, this was not the case. A January 1855 *Observer* article shows that Owen actually had no idea about the Crystal Palace *Megalosaurus* shoulder hump until the models were completed and installed, and the absence of the *Altispinax* bones or a shoulder hump in Owen's 1854 illustration confirms that he was not envisaging the Crystal Palace *Megalosaurus* this way. As reported by the *Observer*:

We believe it was Professor Owen who, sometime afterwards, seeing the animal *in situ*, asked Mr. Hawkins his authority for this additional mound of flesh, and Mr. Hawkins submitted his theory [of head support], and the reply of the learned professor was, that it was perfectly feasible, and that, at all events, no one could say that the bump was not there. *Si non e vero e bentrovato* ['even if it is not true, it is well conceived'].

THE OBSERVER, 7 JANUARY 1855, P. 4

Of note here is that Owen did not mention *Altispinax*, despite the kudos it would have earned both him and the project to do so. Post-1854 boasting from Hawkins also confirms the fact that *Altispinax* was not referenced in the Crystal Palace *Megalosaurus*, as he used its discovery and Owen's interpretation as evidence of his sharp skills for anatomical prediction (Secord, 2004). Owen also gave his 1842 size predictions for *Megalosaurus* in his 1854 guidebook rather than the new estimates he would publish soon after. We cannot be certain if Owen was denying Hawkins and the Crystal Palace Company new data on *Megalosaurus* or if he had simply not produced it, but it is undoubtedly curious that he didn't use his new interpretations to vindicate the project he ostensibly consulted on.

Fig. 7.18 *Megalosaurus bucklandii* as we understand it today: a large theropod (*c.* 7m long) that once roamed Jurassic Britain. The prey species here is *Cetiosaurus oxoniensis*, one of the first sauropod (long-necked) dinosaurs known to science. *Cetiosaurus* was known to Owen and Hawkins when working on the Crystal Palace models, but it was very poorly understood and could not justify even a speculative model. In hindsight this is quite disappointing: a half-speculative Victorian take on a sauropod would surely have been a sight to behold.

The Crystal Palace *Megalosaurus* is restored with its mouth slightly ajar, and it's clear from both this model and associated Hawkins' artwork that the teeth were imagined to occupy a lipless, crocodylian-like mouth. Today, the subject of dental exposure in dinosaurs and other fossil species is a hot topic among palaeoartists, although the majority of informed scientific opinion suggests that predatory dinosaurs bore lizard-like lips (*see* Witton, 2018 for a review). Hawkins' take was in line with general ideas of dinosaur palaeobiology in the 1850s, however, as the socketed nature of dinosaur teeth, and details of *Megalosaurus* jaws, were often favourably compared with those of crocodylians. It seems Hawkins took this interpretation literally in his representation.

The skin texture of the Crystal Palace *Megalosaurus* is unusual among the reptilian models of the Geological Court for lacking obvious scales. Instead, it is deeply fissured and wrinkled in a way that recalls elephant hide. Similar renditions are seen across other Hawkinsian takes on *Megalosaurus* from the mid- and late 1800s. The decision to not depict individual scales was actually quite perceptive, as the eventual discovery of dinosaur skin would reveal the surprisingly small nature of their scales (typically less than a centimetre across), but the thought process behind Hawkins' decision to use this skin texture is not clear. A generous interpretation is of Hawkins predicting that dinosaurs might be atypical reptiles, a fact later vindicated by their skeletal proportions and the discovery of diverse skin types and feathering. Alternatively, perhaps he simply wanted a skin type to distinguish the *Megalosaurus* from the other dinosaurs?

The basic shape of predatory dinosaurs is well known today, even among those with no interest in prehistoric life, and it is thus widely recognized that the genuine *Megalosaurus* was a rather differently shaped animal to that imagined by Owen and Hawkins. Indeed, the fact that predatory dinosaurs were bipedal animals with short arms was obvious within fifteen years of the Geological Court opening (e.g. Cope, 1867, 1869). But much of our knowledge of *Megalosaurus* is still based on the same bones Hawkins worked on in the 1850s: it remains a relatively poorly known dinosaur where proportions and size can only be extrapolated from better known relatives.

Today, *Megalosaurus* is classified as a megalosaurid, a robust-bodied, long-snouted group of predatory dinosaurs from the same evolutionary branch as the famous spinosaurids (e.g. *Spinosaurus*, *Baryonyx*) and allosauroids (*Allosaurus*, *Carcharodontosaurus*). This places it in a region of predatory dinosaur evolution that was still some distance away from the origin of birds, though some aspects of avian anatomy, such as bipedality, details of forelimb function and the extensive air sac system throughout the head, neck and chest, were already apparent in these dinosaurs. We have enough of *Megalosaurus* to predict it was a large-headed creature with a robust, three-fingered hand, and likely similar to the better-known US and Portuguese *Torvosaurus* in general form. Benson (2010) predicted a body length of 6–7m, making it a mid-sized theropod dinosaur overall, but large for the Middle Jurassic world it lived in. In evolutionary terms, *Megalosaurus* is not included within a group incontrovertibly known to have possessed proto-feathers or true feathers, but we also lack substantial skin data for this branch of theropod evolution and feathered skin cannot be entirely ruled out. This said, the scraps of skin we have from *Megalosaurus* relatives are scaly (as seen in the allosauroids *Allosaurus* and *Concovenator* – Pinegar *et al.*, 2003; Ortega *et al.*, 2010), and evidence proposed for feathers or fibres (purported feather anchoring structures along the bones of the forearm) are controversial (Foth *et al.*, 2014).

CHAPTER 8

The sculptures: *'Teleosaurus chapmani'*

With most denizens of the Secondary Island representing unusual, exotic fossil forms, the two *Teleosaurus* sculptures situated close to the *Megalosaurus*, surrounded by blocks of Oolitic limestone, bring a sense of relative familiarity to the middle region of the island. Looking like very large gharials basking on the banks of the Tidal Lake (*see* Fig. 8.1), they act as a grounding agent among the increasingly alien beings encountered by visitors walking through the Geological Court. Like *Megaloceros*, they are visually close enough to modern animals that they offer continuity between Deep Time and the modern day, but still offer a prehistoric twist on a contemporary species.

Constructed with concrete bodies and metal teeth, the 8m-long *Teleosaurus* sculptures are posed with elevated heads but resting torsos and tails. Their slender, elongate jaws are their most striking feature but should not distract from the impressive details worked into their bodies, especially their intricately carved skin, which rivals any other feature of the Geological Court in craftsmanship. The models we see today have been heavily reconstructed after periodical damage that has, at times, seen entire heads and jaws broken apart, tails snapped and extensive cracking spread across their skin. Now painted a leafy green colour, the original paintwork

Fig. 8.1 The *Teleosaurus* – probably '*T. chapmanni*', and thus today regarded as *Macrospondylus bollensis* – pair on the overgrown banks of the Tidal Lake in 2018. Shown here after recent conservation work and thus looking especially well-presented, the waters of the Tidal Lake are usually level with the base of these sculptures.

has proven difficult to reconstruct owing to moisture-accelerated decay.

Unlike the other fossil reptiles of the Geological Court, teleosauroids had been known for almost a full century before Hawkins recreated them in south London. The first complete teleosauroid skeletons, excavated in 1758 and 1791, pre-dated the development of palaeontological science by several decades and received understandably naive interpretations from

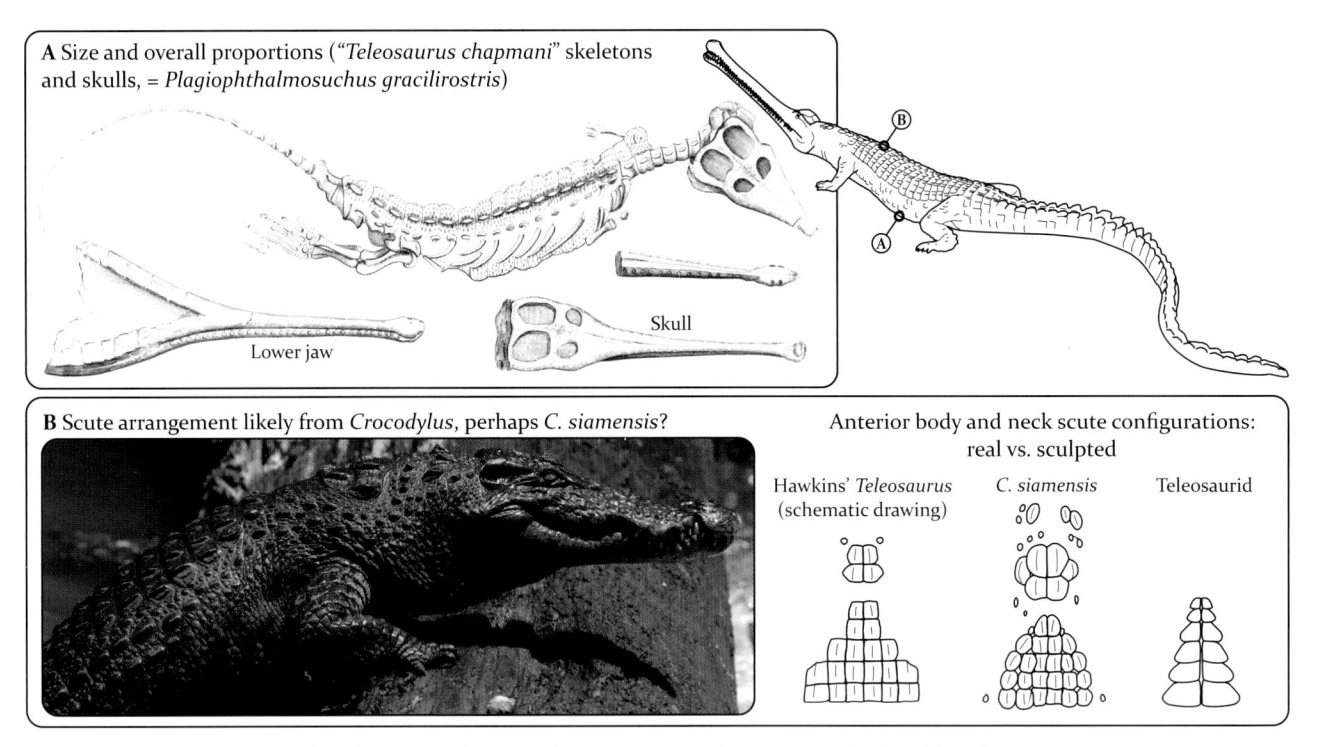

Fig. 8.2 Reference material for the *Teleosaurus* sculptures. '*Teleosaurus chapmani*' skeleton from Buckland (1836), *C. siamensis* scute configuration from Ross and Mayer (1983).

A Size and overall proportions ("*Teleosaurus chapmani*" skeletons and skulls, = *Plagiophthalmosuchus gracilirostris*)

Lower jaw — Skull

B Scute arrangement likely from *Crocodylus*, perhaps *C. siamensis*?

Anterior body and neck scute configurations: real vs. sculpted

Hawkins' *Teleosaurus* (schematic drawing) — *C. siamensis* — Teleosaurid

contemporary scholars. Did these fossils represent ancient alligators or gharials, or even dolphins? Nineteenth-century scientists were better able to place teleosauroids as ancient relatives of crocodylians and they proved popular enough to enter the Victorian palaeoart canon early on, at least by 1830 (Fig. 1.8A).

Their inclusion at Crystal Palace was thus almost a given, but exactly what species of teleosauroid is represented in the Geological Court was not made obvious in contemporary accounts. Owen's (1854) Crystal Palace guidebook implicates an identity of '*Teleosaurus chapmani*' (sometimes spelt '*chapmanni*') via his overview of teleosauroid discoveries from Oolite deposits of Whitby (today the Whitby Mudstone Formation, Lias Group, of the Yorkshire coast, UK). This species was the first named teleosauroid from this unit and was represented by skeletons complete enough to potentially inform Hawkins' whole-body life reconstruction. Other species of *Teleosaurus* had already been named by 1854, however, as had several for another teleosauroid genus, *Steneosaurus*. This was the beginning of a torturous taxonomic history that complicates any efforts to understand early interpretations of teleosauroids – a situation only recently resolved by Michela Johnson and colleagues (2020).

Accordingly, while '*T. chapmani*' is a likely source for Hawkins' reconstruction and matches several important characteristics of the sculptures (e.g. its large size, eye sockets facing outward and upwards – *see* Buckland 1836 and Owen 1842 for historic context) – we cannot verify this as certain. If '*T. chapmani*' is the true identity, then our modern classification of the sculptures – after much taxonomic unwinding – is *Macrospondylus bollensis* (Benton and Taylor, 1984; Johnson *et al.*, 2020). The name *Teleosaurus* remains valid but now pertains exclusively to a French species, *T. cadomensis*, that is not evidenced as being referenced at Crystal Palace.

Teleosauroids, especially relatively early members of the group like *Macrospondylus*, were outwardly similar to modern crocodylians in appearance and probable behaviour, so *Teleosaurus* was probably not the most challenging extinct reptile for Hawkins to reconstruct. Proportions from relatively complete specimens were faithfully reproduced to create sculptures with long, narrow jaws, short limbs, long bodies and long, powerful tails. The depiction of both animals as basking on a shoreline are still consistent with predictions of *Macrospondylus* habits. Although teleosauroids were members of the Thalattosuchia, a group of marine Jurassic and Cretaceous crocodylomorphs, they were among the first representatives of this lineage and not as committed to life in seas and oceans as later species, such as the metriorhynchids. Despite resembling gharials, in ecology

Fig. 8.3 Details of the Crystal Palace *Teleosaurus* (2013 and 2018). Note, among other features, the amazingly detailed skin.

Macrospondylus may have lived more like semi-marine crocodiles of modern times, such as the saltwater crocodile *Crocodylus porosus* (Johnson *et al.*, 2020). In not being specialized for an open water existence, it is likely that *Macrospondylus* and similar grade teleosaurs left water to rest in terrestrial settings, as depicted at Crystal Palace.

Even from afar, the level of detail worked into Hawkins' *Teleosaurus* is evident. Hundreds of metal teeth line the jaws, crocodylian-like cracked skin covers their faces, and hundreds of scales of different shapes tessellate over their bodies. This is basically correct to some aspects of teleosauroid life appearance, with fossilized skin tissues of *Macrospondylus* showing an essentially crocodylian-like scale morphology (Spindler *et al.*, 2021). But while impressively executed, Hawkins and the Crystal Palace team erred in some aspects of depicting teleosauroid skin, evidently basing their scale arrangement wholly on modern crocodylians rather than replicating the teleosauroid armour seen in fossils (*see* Fig. 8.2).

As depicted, the Geological Court *Teleosaurus* have numerous bands of armour on their upper surfaces formed of up to six keeled scales. They are a very close match for those of living crocodylians, with details of spacing and scute shape particularly matching those of *Crocodylus* crocodiles (the current models show greatest similarity to *C. siamensis*, but given their history of restoration this may not be true of their original form). Teleosauroid armour, and thus scalation patterns, were very different to those of *Crocodylus*,

Fig. 157. Le Téléosaure et l'Hyléosaure.

Fig. 8.4 Although terrifically detailed, the scute arrangement in Hawkins' *Teleosaurus* was inconsistent with the known reality of teleosauroid fossils. Contemporary artists were producing more accurate art of these creatures: this 1863 piece by Edward Riou not only captures the armour of the back, but shows a dead, upturned *Teleosaurus* specifically to show the densely tessellated osteoderms of the belly region (from Figuier, 1863).

however. Near-complete fossil skeletons, known since the mid-1700s, show that these reptiles had two broad rows of overlapping, ovate scutes extending from the base of their necks to the mid-region of their tails, while their bellies were covered in a large sheet of bone comprised of tessellating smaller elements. This almost turtle-like configuration is entirely absent in the Crystal Palace *Teleosaurus*, despite it being well known to Victorian anatomists, including Owen, who wrote at length about teleosauroid armour in 1841, as well as artists. Modern crocodylians also seem to have erroneously informed the posterior region of the head, which lacks the long, broad region for jaw muscle housing seen in teleosauroid skulls.

The result are sculptures which blend crocodylian and teleosauroid anatomy. They could, given the state of teleosauroid science in 1854, have thus been more accurate to fossil data. Hawkins returned to *Teleosaurus* in later artwork (*see* Chapter 12) and perpetuated these same errors even as

his contemporaries produced more authentic takes on their anatomy, demonstrating a long-term unawareness of their actual life appearance. Given that *Teleosaurus* was one of the few species restored at Crystal Palace where skin data was known (the only other being *Ichthyosaurus*) it is curious that more efforts were not made to accurately reflect it, especially as some contemporary scientists felt putting any skin detailing on palaeoartworks was pushing our reconstruction abilities too far (*see* Chapter 12). Curiously, Owen made no mention of this error in his 1854 guide, despite being unafraid to draw attention to other mistakes in the display. Perhaps, as with some other sculptures, he did not know of this reconstruction choice, or was he avoiding drawing attention to a mistake that he, as project consultant, could have corrected?

Despite the skew towards crocodylian rather than teleosauroid anatomy in some areas, Hawkins' restorations of *Teleosaurus* are not too far off how we restore these animals

Fig. 8.5 *Macrospondylus bollensis*, the reality of '*Teleosaurus chapmani*', as we might restore it today. As a semi-marine thalattosuchian, *Macrospondylus* was probably still capable of moving around on land, although much of its time would have been spent in the sea.

today (*see* Fig. 8.5). Our largest restorative distinction concerns an appreciation for teleosauroids as unique animals in their own right, and not simply prehistoric versions of modern crocodiles. Modern palaeoartists are accordingly generally more attentive with replicating their proportions, armament and other specifics than Hawkins was in the early 1850s – although living crocodylians thinly disguised as prehistoric species still make it into plenty of modern palaeoart.

Indeed, we are currently experiencing a revived academic interest in teleosauroids after a century of little detailed study, and much is being learned about the taxonomic and ecological diversity they achieved throughout their Jurassic evolution (e.g. Johnson *et al.*, 2020). Teleosauroids are now known to have lived as both generalist and specialist predators in freshwater and marine habitats across much of the world, with some even being adapted to prolonged terrestrial existence. Within this broader group, we can place the Crystal Palace *Macrospondylus* as an early member of the machimosaurids: large and powerful semi-marine predators. Sometimes exceeding 5m in length (Young *et al.*, 2016), *Macrospondylus* likely employed various forms of ambushing tactics to capture aquatic and terrestrial prey in a manner reminiscent of living crocodylians. Recent analysis of rare *Macrospondylus* soft-tissue remains show small, crocodylian-like, square-shaped scales on their legs, and hint at the possibility of paddle-like tissues on their feet like those of gharials and more aquatic *Crocodylus* species (Spindler *et al.*, 2021).

CHAPTER 9

The sculptures: enaliosaurs

After the dinosaurs, the Geological Court's Jurassic marine reptiles are probably the most famous prehistoric residents of Crystal Palace Park. Six models of plesiosaurs and ichthyosaurs are situated around the water margins and small landmasses that comprise the southwest tail of the Secondary Island, some in clear view of public walkways and others partially hidden by the landscaping. In the nineteenth century plesiosaurs and ichthyosaurs were classified together in the group 'Enaliosauria', and it's under this title that Owen presented these sculptures in his 1854 guidebook. This term has since fallen out of use but recent work on marine reptile relationships has revived the idea that plesiosaurs and ichthyosaurs may be close kin: the concept of an enaliosaur group may soon make a comeback.

Unlike the *Mosasaurus*, most of the Crystal Palace enaliosaurs were restored in entirety such that, at the simulated 'low tide' of the Tidal Lake, their whole bodies could be seen (*Ichthyosaurus communis* being the only exception, *see* below). This reflects historic beliefs that enaliosaurs were amphibious creatures rather than, as we imagine them today, fully marine animals incapable of effective land locomotion. Although their appearance is not entirely concordant with modern views, most aspects of the enaliosaur restorations have stood up well against the advancement of scientific knowledge, a status likely reflecting the availability of complete ichthyosaur and plesiosaur skeletons, and even soft-tissue remains, to inform their reconstruction. Their enduring scientific quality is a good demonstration of how Hawkins was capable of restoring strange, unfamiliar fossil reptiles with the same scientific quality as the more familiar fossil mammals when provided with adequate osteological data.

The decision to include Jurassic marine reptiles in the Geological Court would have required little thought. From the early 1800s onwards many exceptional fossil skeletons of these animals had been discovered along the famous 'Jurassic Coast' of Dorset, UK, and they had become a sensation among Victorian scholars. Giant, ferocious ichthyosaurs and writhing, snake-necked plesiosaurs were the first icons of palaeoart: frequently illustrated, immediately recognizable and invariably dominating whatever composition they were set in. Many of the most famous and influential palaeo-artworks of the nineteenth century – de la Beche's *Duria Antiquior* (1830, *see* Fig. 1.8A), Martin's *The Sea-Dragons as they lived* (*see* Fig. 1.9C), Riou's *Ideal Scene of the Lias period* (1863, *see* Fig. 9.2) – were constructed around these animals. Although their celebrity has waned somewhat today (eclipsed, as with most fossil animals, by dinosaurs), animals like *Temnodontosaurus* and *Plesiosaurus* were the original stars of palaeoart, and there was little chance that they would be omitted from the Geological Court.

Fig. 9.1 Enaliosaurs of Crystal Palace, seen in 2015: from left to right, *Plesiosaurus dolichodeirus* and *'Ichthyosaurus'* (= *Temnodontosaurus*) *platyodon* and *'Plesiosaurus'* (= *Thalassiodracon*) *hawkinsi*. These Jurassic marine reptiles occupy a considerable length of the Secondary Island shore and, collectively, are among the most impressive sets of sculptures in the Geological Court.

Fig. 9.2 Edward Riou's famous scene of a battling ichthyosaur and plesiosaur, from Louis Figuier's 1863 book, *La Terre avant le Déluge*. Riou's much copied scene is very typical of nineteenth-century depictions of Jurassic marine reptiles, as were the Crystal Palace models.

Ichthyosaurs

Three moderate- to large-sized ichthyosaur species are featured at Crystal Palace: *Ichthyosaurus communis*, *T. platyodon* and *T. tenuirostris*. Only one – *I. communis* – remains in the genus *Ichthyosaurus* today, it being realized that the other species are anatomically quite different from this classic taxon. They have since been reclassified as *Temnodontosaurus platyodon*, the last of the giant ichthyosaurs, and the large-paddled, slender-snouted *Leptonectes tenuirostris*.

Insights into how these models were built are provided by an illustration of a complete ichthyosaur in Hawkins' workshop (*see* Fig. 3.21A) and a photograph of the developing Geological Court showing a disassembled ichthyosaur in the background (*see* Fig. 3.14). Whether the workshop ichthyosaur represented a clay sculpture or completed model is not clear, but the latter photograph shows that at least some of the concrete ichthyosaurs were built in sections. Each model has required a fair degree of maintenance thanks to the formation of cracks in their skin, weathered flippers, broken jaws and snapped tail fins. Investigations into the *Temnodontosaurus* jaw have revealed a long history of unorthodox conservation work where car body filler and even plasticine have been used as repair agents. At times, freeze-thaw weathering has degraded their flippers to such an extent that they've had to be rebuilt almost entirely. Their situation

Fig. 9.3 The Crystal Palace ichthyosaurs. **A)** *Ichthyosaurus communis* (2018); **B)** '*Ichthyosaurus*' (= *Leptonectes*) *tenuirostris* (2021); **C)** '*Ichthyosaurus*' (= *Temnodontosaurus*) *platyodon* (2011).

Fig. 9.4 Details of the Crystal palace ichthyosaur sculptures. (2013 and 2021)

in water has accelerated decay of their paint layers, such that their original colours cannot be determined even from careful analysis of paint flecks (Hirst Conservation, 2000).

Anatomically speaking, the ichthyosaur sculptures are broadly similar. Though differing in proportion and size, each was restored as a shark-shaped animal with a flexible tail, ornate flippers and dolphin-like facial features. *Temnodontosaurus* is the largest and most conspicuous of the three, with *Leptonectes* and *Ichthyosaurus* set further back from the viewing path. The latter shares a distinction with *Mosasaurus* in being only partly restored: the mid-length of the body sinks into the ground such that only the shoulders, head and tail tip are visible. But unlike *Mosasaurus*, there is no question as to whether incomplete fossils were behind this reconstructive decision: the Crystal Palace ichthyosaurs were all based on complete skeletons from Jurassic deposits of Dorset and Somerset that answered all questions of basic proportions and size. As remarked by Owen:

> Of no extinct reptile are the materials for a complete and exact restoration more abundant and satisfactory than of the *Ichthyosaurus* they [sic] plainly show that its general external figure must have been that of a huge predatory abdominal fish, with a longer tail, and a smaller tail-fin: scale-less, moreover, and covered by a smooth, or finely wrinkled skin analogous to that of the whale tribe.
>
> OWEN, 1854, P. 28

This excellent knowledge of ichthyosaur form had accumulated rapidly. Ichthyosaurs were first discovered by Joseph and his famous fossil-hunting sister Mary Anning in 1811–12, when a complete *Temnodontosaurus platyodon* skull and associated neck vertebrae were discovered in the Lias Group of the Dorset coast. From this point, a huge number of complete and articulated ichthyosaur specimens were exhumed in various UK localities, including many Liassic examples by Mary Anning. As alluded to by Owen, some specimens even had scraps of fossilized soft tissue clinging to their bones, hinting at their scale-less, smooth skin (Buckland, 1836). Details of their flipper shapes were also evidenced in fossil remains and Owen was able to deduce that a soft-tissue fin was situated at the end of the tail. His deduction, however, was the right conclusion reached by

A Proportions and size ("*Ichthyosaurus*" (= *Temnodontosaurus*) *platyodon* skeletons)

B Facial anatomy (3D "*Ichthyosaurus*" (= *Temnodontosaurus*) *platyodon* skulls)

"*Ichthyosaurus*" (= *Temnodontosaurus*) *platyodon*

C Proportions and size ("*Ichthyosaurus*" (= *Leptonectes*) *communis* skeleton)

"*Ichthyosaurus*" (= *Leptonectes*) *tenuirostris*

Ichthyosaurus communis

D Proportions and size (*Ichthyosaurus communis* skeletons)

E Body skin (Ichthyosaur fossil soft-tissues)

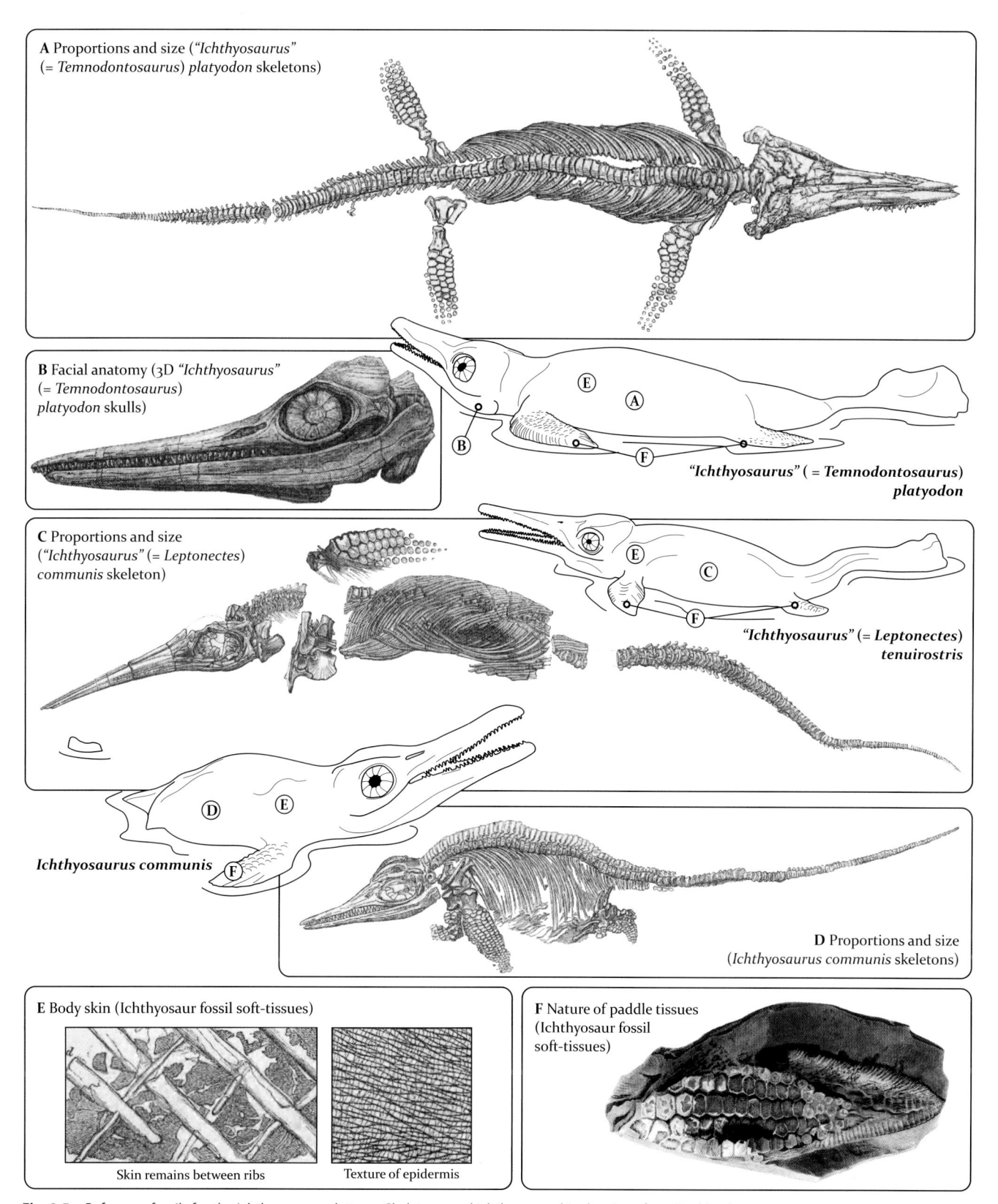

Skin remains between ribs Texture of epidermis

F Nature of paddle tissues (Ichthyosaur fossil soft-tissues)

Fig. 9.5 Reference fossils for the ichthyosaur sculptures. Skeletons and ichthyosaur skin drawings from Buckland (1836), flipper from Owen (1841).

Chapter 9 – The sculptures: enaliosaurs   **129**

flawed logic. Owen assumed that the downward bends seen in all well-preserved ichthyosaur tails were fractures to the spine caused by fin decay (Owen, 1840b) and not – as we now know – genuine features of the ichthyosaur spinal column, where the tail turns down to form the bottom half of a tall, bilobed tail fin. This explains why the Crystal Palace ichthyosaurs also have straight tails – a common convention in nineteenth-century palaeoart.

Owen's (1841) work on an exceptionally preserved flipper also informed the Geological Court ichthyosaurs (*see* Fig. 9.5F). This Lias specimen from Leicestershire showed the full extent of the soft tissues surrounding the flipper bones and revealed a much broader swimming organ than the underlying skeleton alone. It also demonstrated details that Owen interpreted as banded webbing around the flipper margins and a scaly central region, which he described as combining elements of crocodylian and turtle limb anatomy. This configuration was exactly replicated by Hawkins and is especially obvious on the *Temnodontosaurus* (*see* Figs 9.4–9.5).

Details of ichthyosaur eye skeletons were also faithfully recreated. As in most reptiles (including birds), ichthyosaur eyes were supported by sclerotic rings: a circle of bony plates that wrap around the front of the eye. Ichthyosaur sclerotic rings were enormous, indicating that they had huge eyes – among the largest to have ever evolved – and their eye plates were thought by some (Buckland, 1836) to have roles beyond supporting eye tissues: perhaps they also protected their owner's eyes from rapid diving? A January 1854 *Observer* article adds further detail, suggesting that such bones were also 'semi-transparent'. These concepts of armoured eyes covered in semi-transparent bone are probably why Hawkins' ichthyosaurs show the entirety of their eye skeleton rather than, as would be more accurate, covering all of the sclerotic ring with flesh to only reveal the central portion of the eyeball (i.e. the part not covered by bone). This mistake is still surprisingly prevalent in modern ichthyosaur palaeoart, even when modern reptiles – especially raptorial birds such as owls, hawks, eagles and so on – show that even large sclerotic rings are entirely undetectable in fully-fleshed animal faces.

Although ichthyosaur fossils provided Hawkins with an atypically good amount of information for his sculptures, he was seemingly not solely guided by them. Details of the ichthyosaur faces seemingly reflect the facial anatomy of whales and dolphins with their sweeping grooves and generous amount of soft tissue, including full lips along their jaws. This seems entirely reasonable given what we know of ichthyosaur skulls and the relationship between jaw bone surfaces and facial features.

Much about the Victorian vision of ichthyosaurs remains generally accurate today – far more so, in fact, than for most fossil reptiles reconstructed during the 1800s. We still restore ichthyosaurs largely as they are shown at Crystal Palace, albeit with some additional guidance and confidence from fossils showing ichthyosaur body outlines and soft tissues discovered from the late nineteenth century onwards. These fossils remain our best insights into ichthyosaur life appearance and they have continued to yield exciting insights into their soft-tissue anatomy (e.g. Lindgren *et al*., 2018). Recent revelations include, for instance, that Jurassic ichthyosaurs also had dorsal fins and tall, crescent-shaped tail fins, as well as generous layers of blubber-like tissue that streamlined their bodies more than is expressed in the relatively sculpted, muscular-looking Crystal Palace versions. We also now realize that the great amount of tail flexion shown in nineteenth century ichthyosaur art is overstated. The three Crystal Palace species likely had varied capacity for tail flexion in life, with *Temnodontosaurus* and *Leptonectes* probably having more flexible tails than the relatively thunniform ('tuna-like') *Ichthyosaurus communis*, but none had the ability to attain the eel-like tail curvature expressed in the Geological Court.

Our final, and most obvious, distinction with Victorian ichthyosaur concepts is that, today, we do not restore them crawling around in shallow water or basking on land. Such depictions were established well before the Crystal Palace (e.g. Hawkins, 1834). They reflected opinions that the shoulder girdle of ichthyosaurs was somewhat like that of the platypus, and thus much more robust than those of whales or dolphins, and therefore capable of supporting their weight on land (Owen, 1854). This view persisted for much of the nineteenth century (McGowan and Motani, 2003) and Hawkins' basking ichthyosaurs would not have seemed unusual to scholars of the time. Later discoveries, including the exceptional specimens with whole body outlines discussed above, showed how well adapted ichthyosaurs were to marine life and, conversely, how ill-suited they were to moving around on land. Curiously, the 1846 discovery that ichthyosaurs gave birth to live young (Pearce, 1846), and thus did not crawl ashore to lay eggs like turtles, did little to dissuade Owen and his colleagues from their views that ichthyosaurs ventured onto land to sleep, rest or even procreate (Owen, 1854).

The individual Geological Court ichthyosaur species are generally so similar that the outline given above accounts for much of their scientific and artistic background. Some specific points are worth exploring for the individual sculptures, however.

Ichthyosaurus communis

If following the Geological Court pathways from geologically youngest to oldest strata, the first ichthyosaur encountered is also the hardest to fully appreciate: *Ichthyosaurus communis*. This sculpture, the mid-sized (though still enormous) of the three at 7–8m long, is located relatively far back from the pathways surrounding the Tertiary Island, such that much of its anatomy is hidden from view (*see* Figs 9.3–9.4). From this angle, it is difficult to observe that this reconstruction is deliberately incomplete: its torso plunges into the ground in a way that, when surrounded by water, implies the animal is hauling itself ashore to bask. A small portion of low, rectangular tail fin was also crafted, though this was surely only visible to the most eagle-eyed visitors, or anyone visiting the Secondary Island directly.

Today, *I. communis* is regarded as one of the most commonly found Early Jurassic ichthyosaurs. Its remains are found in latest Triassic and early Jurassic sites around the UK but are especially common in Lyme Regis, Dorset, and Street, Somerset (McGowan and Motani, 2003). Its remains are so abundant that we have a record of its growth from young juveniles to fully grown adults, as well as a good insight into its diet of fish and the squid-like belemnites (Lomax *et al.*, 2017). Although restored to match Owen's (1854) suggestion that *I. communis* could reach 20 feet (6.1m) in length, refinements to ichthyosaur taxonomy have shown that *I. communis* was actually a very small ichthyosaur of only 2m long. This places it among the smaller end of all marine reptiles and, indeed, makes it only fractionally larger than the smallest ocean-going cetaceans of modern times.

'Ichthyosaurus' tenuirostris

The smallest ichthyosaur sculpture in the Geological Court represents '*I.*' *tenuirostris* – renamed *Leptonectes tenuirostris* in 1996. This species is well known from latest Triassic deposits of Dorset and Somerset, where its remains are relatively abundant even if complete skeletons are rare (McGowan and Motani, 2003). Most specimens show *L. tenuirostris* was relatively small at just 2.5m long, although some specimens hint at much larger, 4m body lengths. Hawkins' 5.8m-long sculpture is thus somewhat oversized compared to these fossils, outsizing them by 25 per cent.

Fig. 9.6 *Ichthyosaurus communis* as we may imagine it today: a small, dolphin-like animal adapted for rapid swimming. Details of its streamlined body and fin shapes are well established from exceptionally preserved fossils.

Fig. 9.7 A modern take on an unusually proportioned ichthyosaur: the long-snouted, overbitten and large-armed *Leptonectes tenuirostris*. These anatomies were documented in the early 1800s but not translated to Hawkins' sculpture.

Leptonectes is shown at Crystal Palace in a fully-exposed basking pose, but is turned away from visitors such that its face is difficult to view (*see* Figs 9.3–9.4). As with *Hylaeosaurus*, the rationale to face this animal away from the public is not clear, although it does show the unusual tail which has a low fin along most of its length, terminating with only a modest vertical expansion at the tip. Conversely, it also hides the distinctively narrow, long snout that typifies leptonectids. This condition was only marginally pronounced in *Leptonectes*, but was taken to an extreme by close relatives such as *Eurhinosaurus* and *Excalibosaurus*. Hawkins' *Leptonectes* possesses an obviously longer, narrower snout than his other ichthyosaurs but, when contrasted with its reference fossils, it does not quite capture the very slender nature of the real jaws, nor the characteristic overbite (*see* Fig. 9.7). Similarly, while Hawkins has shown the fore-flippers as large relative to the hind-flippers, he was overly conservative with their distinction: the forelimbs of *Leptonectes* are actually among the largest in any ichthyosaur, relative to body size.

'Ichthyosaurus' platyodon

Perhaps the most recognizable sculpture of the Geological Court after the dinosaurs, the giant '*Ichthyosaurus*' *platyodon* – today known as *Temnodontosaurus platyodon* – is an imposing, large restoration that represents the third and last ichthyosaur in the marine reptile sequence. It is difficult to appreciate from a distance how large this model is. Roughly 10m long measured from snout to tail tip, and presumably 12m or more if straightened out, it is comparable to the biggest dinosaur sculptures in size. The head alone is 2.3m long and enormously bulky when viewed at close range, and the foreflippers measure a full 1.7m along their midline.

These proportions are apt for *Temnodontosaurus*, which has been regarded since its discovery as one of the largest of all ichthyosaurs. Its maximum size is not reliably known because only one mostly-complete skeleton of *T. platyodon* has been found (Fig. 9.5A) and it represents a mid-sized animal of *c.* 6m long, but 2m-long skulls from the same Early Jurassic beds of Dorset indicate specimens of approximately 9m in length (McGowan, 1996). It is almost certainly the size, and formidable jaws and teeth, of *Temnodontosaurus*

Fig. 9.8 A 2021 take on *Temnodontosaurus platyodon*, a large-bodied ichthyosaur that predated other marine reptiles. The chief difference between this modern take and Hawkins' Crystal Palace model is the presence of a dorsal fin, which had yet to be evidenced for ichthyosaurs in 1854. The prey item in this image is a juvenile rhomaleosaurid.

that saw it briefly attain status as one of the most restored species in early palaeoart, it being the focus of innumerable artworks including the iconic *Duria Antiquior* and its imitators (*see* Fig. 1.8), as well as John Martin's *The Sea Dragons as they lived* (*see* Fig. 1.9C). Often depicted as ravaging other marine reptiles, *Temnodontosaurus* was the first in a long line of gigantic, formidable reptiles that would become iconic fixtures in palaeoart: the *Tyrannosaurus* of its day.

The Crystal Palace *Temnodontosaurus* is the only ichthyosaur fully visible from public footpaths. It is depicted, like the others, as basking, albeit on a flat bank that allows the entire base of the sculpture to submerge in water (*see* Figs 9.3–9.4). As noted by Owen and captured faithfully by Hawkins, the head of this species is proportionally large and robust, befitting that of an ichthyosaur long-considered an apex predator of Jurassic seas. Hawkins also reproduced the equally proportioned fore- and hind-flippers accordingly, and speculated – probably correctly, based on complete *Temnodontosaurus* remains – a relatively large and tall tail fin. Of the three ichthyosaurs, Hawkins' *Temnodontosaurus* can be regarded as being the most authentic with respect to modern science, as well as the most imposing of his marine reptile restorations.

Plesiosaurs

The neighbours of the Crystal Palace ichthyosaurs are the plesiosaurs, three sculptures looking something like swan-necked seals, arranged around the banks of the Secondary Island. They are dwarfed by the larger, bulky ichthyosaurs but maintain an arresting presence from their bizarre appearance. Though familiar to us today, the plesiosaur body plan was regarded as the most aberrant known among fossil animals in the mid-1800s, such that the discovery of a complete skeleton at the hands of Mary Anning in 1823 was initially met with incredulity. Some academics went so far as to assume the fossil was faked (Cadbury, 2000; Emling, 2009).

The authenticity of this specimen was soon established but the four-flippered, long-necked body plan it presented flummoxed nineteenth-century scholars as much as it excited them. Such was the interest in plesiosaurs that Anning was paid £110 (£13,390 adjusted for inflation) for her complete plesiosaur fossil, the highest price yet paid for a fossil reptile specimen (Emling, 2009). It seemed unbelievable that so many contrasting anatomies could be present in one animal, which geologist William Buckland summarized as:

Fig. 9.9 The Crystal Palace plesiosaurs in various states of restoration. **A)** *'Plesiosaurus' macrocephalus* in 2020; **B)** *Plesiosaurus dolichodeirus* (2013); **C)** *'Plesiosaurus'* (= *Thalassiodracon*) *hawkinsii* (2013).

> It is of the *Plesiosaurus*, that Cuvier asserts the structure to have been the most heteroclite, and its characters altogether the most monstrous, that have been yet found amid the ruins of a former world. To the head of a lizard it united the body of a serpent, a trunk and tail having the proportion of an ordinary quadruped, the ribs of a chameleon, and the paddles of a whale.
>
> BUCKLAND, 1836, PP. 202–03

Plesiosaurs soon became familiar to nineteenth-century palaeontologists by way of additional specimens, including many more complete skeletons from the southern UK, but their enigmatic nature has never truly evaporated. Although we now appreciate plesiosaurs less as monstrous animals designed by committee and more as unique, spectacular examples of marine reptiles without satisfactory modern analogues, many details of plesiosaur anatomy and ecology remain controversial. How did they use their unique configuration of four-flippers for propulsion (Liu *et al.*, 2015; Muscutt *et al.*, 2017)? What was the ecological significance of their long neck (Noè *et al.*, 2017)? Many of us interested in palaeontology are so used to plesiosaurs that we've become overly familiar and blasé about their unusual body shape, but they surely remain one of the most peculiar groups of Mesozoic marine reptiles.

As with the ichthyosaurs, the three plesiosaurs of Crystal Palace represent three different species, all of which stem from the latest Triassic and Early Jurassic of Dorset and Somerset, UK. All three were considered members of the genus *Plesiosaurus*: *Ple. dolichodeirus*, *'Ple.' hawkinsii* and *'Ple.' macrocephalus*. Despite representing different taxa, they represent a relatively homogenous set of sculptures distinguished mostly by subtle body proportions and overall size.

As with all the partially submerged models, centuries of persistent moisture have enhanced the decay of their original paint layers and their original colours have long corroded away. Illustrations of Hawkins' workshop (*see* Fig. 3.21A) and photographs of the half-constructed Geological Court (*see* Fig. 3.15) show that the plesiosaurs were built indoors before being transported to their external sites. Their heads were constructed from metal, presumably reflecting the need for finer detail and lower weight than would be afforded by the concrete, iron and brick forming the torsos, tails and flippers. At times the plesiosaurs have become so weathered by freeze-thaw that their flippers have been rebuilt from virtually nothing. Their necks are also vulnerable to damage and have proved prone to cracking and snapping. At some point in the early or mid-twentieth century the head of *'Ple.' macrocephalus* was lost and a replacement modelled on the *'Ple.' hawkinsii* sculpture was installed in the 1950s. The 1959 film *Restoring Monsters* by British Pathé suggests that the original *macrocephalus* head was lost during the bombing of London during the Second World War: if so, this would be the only recorded instance of the Geological Court being damaged in the Blitz. The remains of the head were subsequently stored in the collections at the Crystal Palace Museum, and were only identified in 2018 by the Friends of Crystal Palace Dinosaurs.

The plesiosaurs are among the smaller reptile sculptures in the park despite Owen's (1854) guidebook suggesting one sculpture is 18ft (5.4m) long: in reality, none exceed 4m. Owen's figure might represent an early concept which was later altered, as it's easy to believe that building creatures

A Proportions and size (*Plesiosaurus dolichodeirus* skeleton)

Plesiosaurus dolichodeirus

B Proportions and size ("*Plesiosaurus*" *macrocephalus* skeleton)

"*Plesiosaurus*" *macrocephalus*

C Flipper articulation (pinnipeds, *Halichoerus grypus* shown)

D Proportions and size ("*Plesiosaurus*" (= *Thalassiodracon*) *hawkinsii* skeleton)

"*Plesiosaurus*" (= *Thalassiodracon*) *hawkinsii*

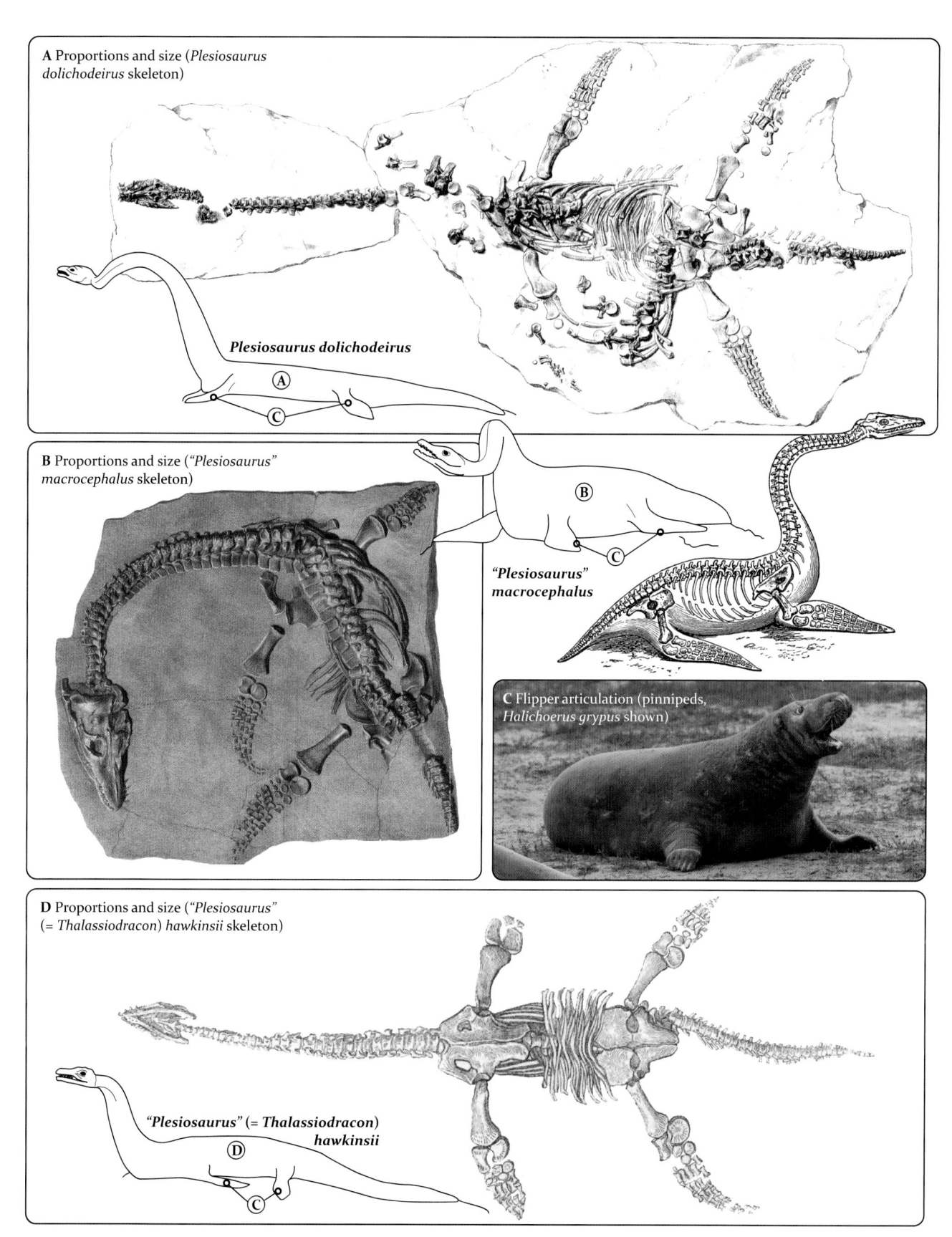

Fig. 9.10 Fossil and living animal references likely consulted for the plesiosaur sculptures. *Plesiosaurus* and *Thalassiodracon* skeletons from Buckland (1836); '*Ple.*' *macrocephalus* fossil from Owen (1840c) and skeletal reconstruction from Owen (1854).

with long, thin necks at gigantic proportions from rudimentary materials would have been a steep engineering challenge for the Geological Court team. This may also explain a confirmed deviation from a more dynamic original composition, where Hawkins imagined his plesiosaurs fighting amongst themselves. As reported by a journalist for the *Hogg's Instructor*:

> Whether Mr Hawkins really intends to carry out the design he showed us sketched on paper, of having two of these monsters represented as in the heat of a deadly feud, their bodies half raised from the ground, their long necks intertwined, and, with glaring eyes, each burying its teeth in the flesh of the other, we can hardly say; he has, however, already completed one which will be scarcely less attractive.
>
> ANONYMOUS, 1854, P. 283

With so little insight into the early development of the Geological Court, it remains uncertain what factors changed Hawkins' vision: perhaps they were practical, accounting for the difficulties of rendering highly poised, wrestling long-necked creatures from iron and concrete? Or could they have been philosophical, avoiding a speculative scene that could be attacked by conservatively minded critics? Or even something else, such as avoiding a tonal clash with the more sedate animals of the rest of the display?

The excellent plesiosaur fossils available to Hawkins allowed him to reconstruct these animals in a form recognizable to us today, although the results were not as insightful as his ichthyosaurs. In part, this was a reflection of the challenges plesiosaurs presented to Victorian scientists and artists. These unusual reptiles were – and still are – simply less intuitive to restore than the likes of ichthyosaurs and fossil mammals. The results were sculptures well-aligned with nineteenth-century theory on plesiosaur life appearance and habits, but with a few idiosyncratic Hawkinsian features that were problematic even for the time. Nevertheless, application of Hawkins' anatomical knowledge means that, for all their issues, they still look like plausible creatures that could exist, had evolution taken a deviant path.

Posed as if resting on the banks of the Secondary Island, the plesiosaurs are arranged as if to have hauled their slender, winding bodies and arching necks out of shallow waters by turtle-like action of their flippers (*see* Figs 9.9, 9.11). As with ichthyosaurs, the strength of the plesiosaurian shoulder girdle was taken as evidence of their ability to move on land (Owen, 1854). Although Owen (1854) compared plesiosaur flippers to those of turtles, Hawkins seems to have made particular reference to the limbs of seals with his depiction of plesiosaurs showing ample fore and aft limb motion as well as obvious elbow and knee joints. Also like seals, his plesiosaurs have smooth and unornamented bodies, excepting some bulges and creases reflecting the underlying muscle and skeletal contours. The tails are characteristically thick and seemingly well-muscled – a contrast to many other nineteenth-century depictions of plesiosaurs. Their slender torsos are shown as being capable of considerable lateral flexion, a feature recalling the lizards that plesiosaurs were considered allied to in the early and mid-1800s. Although difficult to see from afar, Hawkins also imbued the heads of his plesiosaurs with several fine details. They include eyes that are somewhat angled upwards and large teeth set within seemingly lipless jaws, features that are closely observed from plesiosaur skulls.

Each sculpture has unique proportions that somewhat reflect those of their source fossils, although the presence of several measuring errors rank them among Hawkins' less successful works at the Geological Court. His depiction of thin, highly flexible necks was entirely in keeping with contemporary theory, however, as expressed by Owen (1854), quoting marine reptile expert Thomas Hawkins (unrelated to Waterhouse) in his guidebook:

> May it not, therefore, be concluded that [a plesiosaur] swam upon, or near the surface arching back its long neck like a swan, and occasionally darting it down at the fish that happened to float within its reach? It may perhaps have lurked in shoal-water along the coast, concealed among the sea-weed, and, raising its nostrils to a level with the surface from a considerable depth, may have found a secure retreat from the assaults of dangerous enemies; while the length and flexibility of its neck may have compensated for the want of strength in its jaws, and its incapacity for swift motion through the water, by the suddenness and agility of the attack which enabled it to make on every animal fitted for its prey which came within its reach.
>
> HAWKINS, 1841
> (QUOTED IN OWEN, 1854, PP. 33–34)

Both advances in our understanding of plesiosaur palaeobiology as well as mistakes in Hawkins' reconstructions mean our contemporary takes on plesiosaurs differ in many regards from those at Crystal Palace. Most obviously, it is no longer thought that plesiosaurs were capable of any terrestrial locomotion, even if buoyed up in shallow water. Not only is their anatomy entirely ill-suited to movement on land, but we now have good evidence that even their reproduction took place wholly at sea, with mother plesiosaurs giving birth to large calves (O'Keefe and Chiappe, 2011). Their flippers also lacked joints distal to the shoulder and hip, and were thus incapable of flexion along their length. Plesiosaur bodies were broad and stiffened by extensive belly ribs and large limb girdles, and were thus not narrow and flexible as depicted at Crystal Palace. Indeed, the discovery of plesiosaur body outlines and flipper soft-tissues has shown that their body contours exceeded even the generous proportions of their skeletons (Frey *et al.*, 2017), including substantial soft tissue around their tails (perhaps reflecting expanded hindlimb musculature running between the upper leg and tails, as is the case for most reptiles) as well as body-contouring fatty tissues. Their flippers were also augmented with soft-tissue extensions, something Owen would predict a few years after the Crystal Palace opened (Owen, 1861) but would not be confirmed from plesiosaur fossils until the late nineteenth century. Further soft-tissue elaboration is suspected at their tail tips, which likely bore flukes or fins of some kind (Smith, 2013; Sennikov, 2019).

Finally, while the flexibility of plesiosaur necks remains a topic of investigation, details of their neck vertebrae suggest they were incapable of pronounced elevation or the adoption of complex, snake-like poses (Noè *et al.*, 2017). Rather, plesiosaur necks are thought to have been well-muscled and relatively stiff – better suited at forming arcs than snake-like knots – as well as achieving their greatest range of motion below and, to a lesser extent, to the side of their bodies.

As with Hawkins' ichthyosaurs, the Geological Court plesiosaurs are broadly similar in anatomy and appearance, but features of each sculpture are worthy of closer attention.

'Plesiosaurus' macrocephalus

When traversing the Geological Court from youngest to older strata, the first plesiosaur encountered at Crystal Palace is the largest: the 4m-long *'Plesiosaurus' macrocephalus*, from the Early Jurassic Lower Lias of Dorset, UK

Fig. 9.11 Details of the three plesiosaur sculptures (2021).

Fig. 9.12 Modern life reconstruction of *'Plesiosaurus' macrocephalus*. This restoration follows ideas that our only fossil of this species is a juvenile of the 6–7m-long rhomaleosaurid *Thaumatodracon wiedenrothi*. Note the large head and mid-sized neck, features overlooked in the Crystal Palace sculpture.

(*see* Figs 9.9, 9.11). A bulky, hulking body with powerful-looking shoulders tapers into a coiled, slender neck that – thanks to the damage and conservation work discussed above – now terminates in a small head similar to that of *'Plesiosaurus' hawkinsii*. Photographs and illustrations of the original, undamaged sculpture show that the *macrocephalus* once had a larger head, turned somewhat upwards, and with an open mouth. It thus may not have originally stared at park visitors quite as directly as it does today. An attempt to illustrate the sculpture with its original head and open jaws, based on low-quality photographs, is seen in Fig. 9.10.

The head is not, admittedly, as large as might be expected given the proportions of the excellent fossil material available to Hawkins: a well-preserved, essentially complete and articulated skeleton described by Owen (1840c) where the skull is clearly over half the length of the torso (*see* Fig. 9.10). It's possible that the challenges of engineering such a large head on a long neck from heavy components resulted in a compromised reconstruction. The skeletal reconstruction of *macrocephalus* in Owen's (1854) guidebook also sports a small head, but given the other mistakes and oversights in this text, this illustration may be a mislabelled *Plesiosaurus dolichodeirus* (*see* Chapter 10 for another, undoubted error of this nature).

Surprisingly, our modern understanding of *'Ple.' macrocephalus* is not significantly advanced over that achieved by Owen almost 200 years ago. Although scholars have referred several specimens to this species, today we only regard one specimen – the original – as a genuine record of this animal.

The fact this fossil represents a young juvenile has hampered abilities to determine exactly what sort of plesiosaur it is, however: juvenile animals tend to lack the taxonomically diagnostic anatomy of older, adult individuals, so finding their homes in evolutionary trees can be challenging. Furthermore, detailed studies into the anatomy and affinities of *macrocephalus* have been hampered by our only specimen being placed in a wonderful but extremely inaccessible Victorian-era display in London's Natural History Museum. Both heavy vintage glass and elevated mounting positions keep many marine reptile fossils in this exhibit out of research reach and, until our *macrocephalus* material can be properly examined, we will be unable to determine what type of plesiosaur it represents.

Despite these issues, there is universal consensus that *macrocephalus* does not belong to the *Plesiosaurus* genus. As with *'Plesiosaurus' hawkinsii* (below), this referral reflects a Victorian convention of placing virtually all plesiosaur fossils within the genus *Plesiosaurus*. In all likelihood, *macrocephalus* was not even closely related to this genus. The large skull, details of the neck and provenance of the specimen imply a home among the rhomaleosaurids (Smith and Araújo, 2017), potentially as a juvenile individual of the contemporary, 6.5m-long rhomaleosaurid *Thaumatodracon wiedenrothi* (Smith, pers. comm. 2021). Rhomaleosaurids were one of the first major radiations of plesiosaurs, characterized by necks of moderate size, large heads, and body lengths up to 7m. A short-lived group compared to other plesiosaur lineages, they acted as large carnivores in Early and Middle

Fig. 9.13 The classic plesiosaur, *Plesiosaurus dolichodeirus*, as imagined today. Ideas that these animals were capable of rearing their heads and necks from the water, or that they could haul themselves ashore, are now long abandoned. Here, *Plesiosaurus* is scavenging a pterosaur – an assumed favourite foodstuff of plesiosaurs among nineteenth-century artists.

Jurassic seas and were thus a distinctive, important component of Mesozoic marine ecology. If *macrocephalus* was, indeed, a rhomaleosaurid, the Geological Court sculpture represents one of the earliest – if not *the* earliest – attempts to restore one of these important but overlooked types of marine reptile.

Plesiosaurus dolichodeirus

Easily identified by having the longest neck of any plesiosaur in the Geological Court, Hawkins' sculpture of *Plesiosaurus dolichodeirus* is situated immediately in front of *Temnodontosaurus* (see Figs 9.9, 9.11). Proportionally the most accurate of Hawkins' plesiosaurs, if still lacking accurate flipper sizes and being a little on the slender slide, this *c.* 4m-long sculpture would have been guided by complete *Plesiosaurus* skeletons from Lias deposits of Dorset, UK (see Fig. 9.10). It is fractionally longer than the *Plesiosaurus* specimens known to Hawkins and Owen, which hinted at animals approaching 3m long, but it captures the small head, long neck and moderate tail length of this species. Some of this anatomy is difficult to make out from visitor paths as the sculpture is posed facing away from the Crystal Palace mainland. The head, in particular, is difficult to observe.

Today, *Ple. dolichodeirus* is the only valid species of the once expansive genus *Plesiosaurus* and it is recognized as an early member of the Plesiosauroidea: a group which contains most of the long-necked plesiosaurians. Although no longer regarded as an especially long-necked species (the famous Cretaceous elasmosaurids have necks up to 60 per cent of their body lengths), it represented the known extreme of this plesiosaur body plan in the mid-nineteenth century. So distinct was its neck that Owen (1854) simply calls it 'Long Necked *Plesiosaurus*' in *Geology and Inhabitants of the Ancient World* (p. 32).

The functional properties of long plesiosaur necks remain contested among researchers, but the idea promoted by Owen (1854, see quote, above) that these animals acted something like swans, with long necks searching for food below the body, is not entirely at odds with some modern thoughts on plesiosaur lifestyles. Attempts to rationalize the entire long-necked plesiosaur body plan suggest they were not fast-moving animals that chased large prey like the rhomaleosaurids or the large-headed, megapredatory pliosaurids, but that they used their small, lightly built skulls and teeth to catch small prey while swimming above them. We can imagine long-necked plesiosaurs swimming slowly or hovering in the water column, sometimes not far from the seabed, while their necks swept downward in search of small swimming prey and shellfish (Noè *et al.*, 2017). Fossilized gut remains show that these prey items were sometimes sieved straight from underwater sediments (McHenry *et al.*, 2005).

Chapter 9 – The sculptures: enaliosaurs

'Plesiosaurus' hawkinsii

The final plesiosaur of the Geological Court is *'Plesiosaurus' hawkinsii*, a sculpture best distinguished by its small size (*see* Figs 9.9, 9.11). Known today as *Thalassiodracon hawkinsii* and represented by several specimens from rocks straddling the Triassic/Jurassic boundary around the southern UK, the 3.8m-long sculpture hints at the small stature of this species compared to the other plesiosaurs in the Geological Court, but exceeds the real body 2m length of this characteristically small species. At first glance, the sculpture appears to be erroneously short-necked because the neck base continues along the ground, unelevated, from the shoulder region, so that only the anterior half of the neck is lifted. The position of the front flippers, however, shows where the true neck base lies and that the neck length was accurately transcribed from the fossil, as was the relatively long tail. As with *Ple. dolichodeirus*, the flippers are undersized relative to their fossil remains.

Although *Thalassiodracon* looks outwardly like a smaller version of *Plesiosaurus*, modern investigations suggest it was an early member of a third major plesiosaur lineage: the pliosaurids (Benson *et al.*, 2011). These animals are among the most popular plesiosaurs thanks to the charismatic appeal of large or gigantic, big-headed predators like *Liopleurodon*, *Pliosaurus* and *Kronosaurus*, but *Thalassiodracon* represents a very early stage in pliosaurid evolution far from the development of such anatomies. In life, it may have lived in a manner similar to that described above for *Plesiosaurus*, using its long neck and small head to seek diminutive prey species from elevated feeding positions.

Fig. 9.14 One of the smallest plesiosaurs, *Thalassiodracon hawkinsii*, tackling a formidable adversary in a modern life reconstruction. Fossils show that plesiosaurs were muscular, streamlined animals with a layer of fatty tissue covering at least some of their bodies, in contrast to Hawkins' slender reconstructions.

CHAPTER 10

The sculptures: *Labyrinthodon*

The southern extent of the Secondary Island is a low-lying landmass covered by several mid-sized crawling forms: the frog-like '*Labyrinthodon*' and turtle-like *Dicynodon*. This region of the Geological Court was built to represent the Triassic New Red Sandstone and is the last component with palaeontological restorations. Although neither the largest nor the most spectacular sculptures, both '*Labyrinthodon*' and *Dicynodon* are fascinating additions to the Geological Court, representing bold but flawed interpretations of extinct animals that were not well understood by scholars of the 1850s. Their highly fragmentary, confusing fossil remains were the ultimate test of the correlative principles underlying Hawkins' design process and his interpretations were far from accurate. This is not to say that '*Labyrinthodon*' and *Dicynodon* are lacking in Hawkinsian insight and reconstructive nuance, however: they can be viewed as intelligent, reasonable interpretations of fossil data ascribed to these animals, while simultaneously exposing the gulf between the assumed and actual correlative capabilities of Victorian palaeontology.

The three Crystal Palace '*Labyrinthodon*' reconstructions are arranged along the banks of their low island as if emerging from, or returning to, the water. In January 1854 *The Observer* reported that they were Hawkins' 'favourite child, upon which he has bestowed unusual pains, as being of course the ugliest in his assemblage of monsters' (p. 5). They are bulky, frog-like animals in basic form, having disproportionately long hindlimbs, wide faces and squat, rounded bodies with truncated, robust tails. Two species are represented, the largest, smooth-skinned model being '*Labyrinthodon salamandroides*' and the two smaller, warty-skinned models representing '*Labyrinthodon pachygnathus*'. Whether '*Labyrinthodon*' was an amphibian or reptile was the matter of debate in the mid-1800s (Moser and Schoch, 2007) and, perhaps unsurprisingly, Hawkins sided with Owen's preferred amphibian identity.

It should be stressed that the term 'amphibian' is used here in the loosest, most vernacular sense. Although amphibious in lifestyle, the large, robust prehistoric 'amphibian' stock of '*Labyrinthodon*' and its kin were very different creatures from our modern frogs, salamanders and caecilians. The broad group housing '*Labyrinthodon*', known as Temnospondyli, was a major radiation of amphibious limbed vertebrates that enjoyed particular success in the Carboniferous, Permian and Triassic periods. They often bore scales, claws and even body armour, and could be fully terrestrial, only returning to water to reproduce. The term 'amphibian' has a much broader meaning when viewed through the lens of Deep Time than it does in the modern day.

Reasonable records exist for the state of the '*Labyrinthodon*' sculptures over time. Their original colour was yellowish-green (McCarthy and Gilbert, 1994; Hirst Conservation, 2000) and – so far as is recorded in photographs – they have also been painted yellows, browns and greys. They appear to be largely constructed from building materials rather than, as with some of the other smaller models, with metal skin. Perhaps on account of their low, stable construction and inaccessibility, they seem to have survived to the modern day with only moderate repair and rebuilding, although the recreated footprints once accompanying the sculptures have disappeared under unknown circumstances (see below).

Everything about the 'Labyrinthodon' genus was, a product of Richard Owen's intellect, eccentricities and misplaced confidence in anatomical correlation. Even the name was a product of Owen's attitude to the work of others. Originally called *Mastodonsaurus*, Owen felt this name implied comparison with mastodons (the extinct, elephant-like animals of North America) so he renamed it 'Labyrinthodon' after his discovery of the complex, labyrinthine tissues of temnospondyl teeth (Owen, 1842d). He similarly replaced the original two proposed *Mastodonsaurus* species names of *jaegeri* and *giganteus* with his preferred alternative, *salamandroides*. This referenced Owen's correct identification of amphibian features in *Mastodonsaurus* fossils and, on the assumption that these large temnospondyls were the only limbed vertebrates from the New Red Sandstone, he ascribed other fossils from this unit to his giant amphibian, too. Today, we know that Owen was actually combining reptilian and amphibian fossils in his 'Labyrinthodon' inventories but, in his defence, the fossils he was working with were fragments and isolated bones. Moreover, in spite of the low quality fossils he was working with, Owen identified the blend of amphibian and reptile anatomies in these specimens and interpreted them accordingly, regarding 'Labyrinthodon' as some sort of 'sauroid batrachian' (Owen, 1842e, p. 542).

A large, complete skull from Germany was the most entirely known element of this mysterious animal. With this as a reference, Owen was able to identify comparable jaws, teeth and fragmentary skull bones among British strata (Owen, 1842e). Pelvic and limb material from the New Red Sandstone were subsumed into 'Labyrinthodon' too, and Owen further assumed that his giant amphibian was the creator of mysterious footprints from Hildburghausen, Germany and Cheshire, UK (see Fig. 10.2B; Owen, 1842e). These trackways, named *Chirotherium barthi* (often spelt '*Cheirotherium*', but originally formulated as '*Chirotherium*' – see Bowden et al., 2010), recorded a quadrupedal animal with vast footprints but tiny handprints. Combining this inventory of bones and tracks, Owen (1842e) created a detailed description of the giant frog-like creatures that Hawkins would restore at Crystal Palace:

> The cranial fragments correspond in size with those of the head of a Crocodile between six and seven feet in length, but the ilium supports an articular cavity for the reception of the head of the femur... of a

Fig. 10.1 The three 'Labyrinthodon' of Crystal Palace. **A)** the larger, longer-legged and smooth-skinned '*Labyrinthodon salamandroides*' (2015); **B)** and **C)** the two '*Labyrinthodon pachygnathus*' models, one facing water, one facing inland, from the southern extent of the Secondary Island (2018).

Fig. 10.2 The fossils and track data that informed the Geological Court 'Labyrinthodon'. The part-skeletal restoration, part-life reconstruction illustration from Owen (1854) was labelled 'L. salamandroides' in the Geological Court guidebook, but – as can be seen from the fossil inventory shown here – is clearly meant to be 'L. pachygnathus'. *Chiortherium* prints from Buckland (1836); Fossil bones from Owen (1842e) and Plieninger (1844).

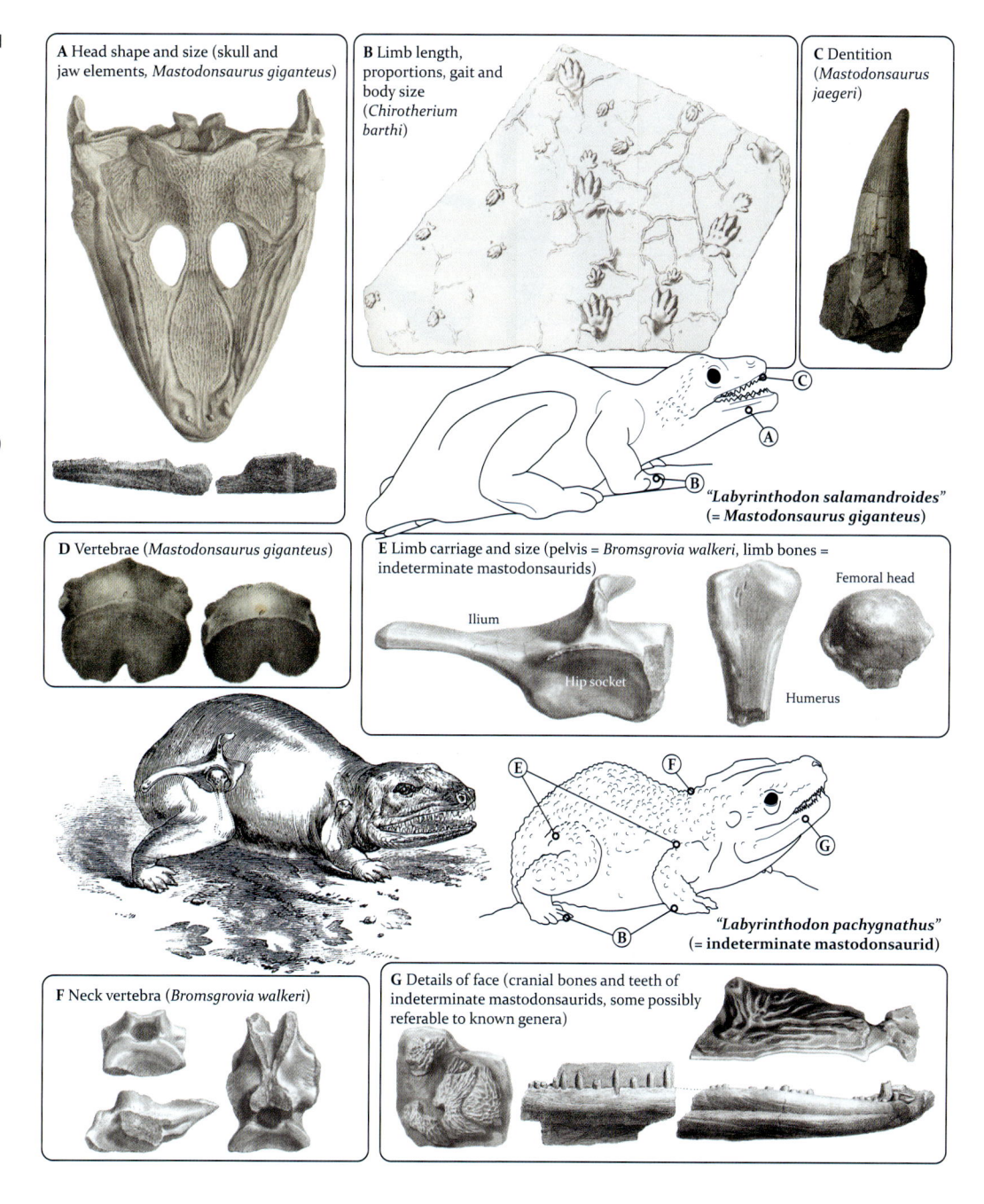

A Head shape and size (skull and jaw elements, *Mastodonsaurus giganteus*)

B Limb length, proportions, gait and body size (*Chirotherium barthi*)

C Dentition (*Mastodonsaurus jaegeri*)

"*Labyrinthodon salamandroides*" (= *Mastodonsaurus giganteus*)

D Vertebrae (*Mastodonsaurus giganteus*)

E Limb carriage and size (pelvis = *Bromsgrovia walkeri*, limb bones = indeterminate mastodonsaurids)

Ilium, Hip socket, Femoral head, Humerus

"*Labyrinthodon pachygnathus*" (= indeterminate mastodonsaurid)

F Neck vertebra (*Bromsgrovia walkeri*)

G Details of face (cranial bones and teeth of indeterminate mastodonsaurids, some possibly referable to known genera)

Crocodile twenty-five feet in length. If both belonged to the same individual, we should have an example of a reptile with hinder extremities of disproportionate magnitude as compared with those of existing Saurians, but which would approximate in this respect… some of the existing anurous Batrachians.

OWEN, 1842E, PP. 534–35

The correlative assembly of Owen's 'sauroid batrachians' seems outwardly similar to his famous and essentially correct *Dinornis* prediction (*see* Chapter 2), but instead actually undermines the principles of anatomical correlation.

Working with older, unfamiliar Triassic animals was not as straightforward as making deductions about the forms of Eocene mammals or recently extinct birds, and Owen's lumping of limbed vertebrate fossils from Triassic rocks into 'Labyrinthodon' saw him create a fictional animal: a chimera of temnospondyl and reptile remains that bore little resemblance to any one of its constituent species.

While Hawkins' Crystal Palace sculptures show loyalty to Owen's concept, other artistic interpretations of his 'Labyrinthodon' were available. Franz Unger's 1851 *Die Urwelt in ihren verschiedenen Bildungsperioden* ('The Primitive World in its Different Periods of Formation') contains earlier efforts

Fig. 10.3 Josef Kuwasseg's illustrations of prehistoric 'amphibians' from Franz Unger's 1851 *Die Urwelt in ihren verschiedenen Bildungsperioden*, showing more salamander-like forms than those of Crystal Palace. **A)** *The Period of the Bunder Sandstone*, showing an unnamed, salamander-like creature, probably 'L. salamandroides'; **B)** *The Period of the Keuper Sandstone*, with a tailed 'L. pachygnathus'.

at restoring Triassic amphibians by Josef Kuwasseg which, even though specifically referencing Owen's work, are more salamander-like in form. Kuwasseg restored two amphibians for Unger's book, an unspecified giant salamander as well as a crouching, long-legged and long-tailed '*Labyrinthodon pachygnathus*'. Although Owen (1854) confidently classified '*Labyrinthodon*' as a frog in *Geology and Inhabitants of the Ancient World* (they are unambiguously listed under 'Batrachia'), he also acknowledged how little guidance artists like Hawkins and Kuwasseg had to work with, noting for '*L. salamandroides*' that:

It is to be understood, however, that, with the exception of the head, the form of the animal is more or less conjectural.

OWEN, 1854, P. 38

Owen would eventually produce his own sketch of *Labyrinthodon pachygnathus* that would be printed in Charles Lyell's 1855 *Manual of Elementary Geology* (see Fig. 10.4B). This famous line sketch is a close match to the frog-like creatures of Crystal Palace, showing some eventual endorsement for Hawkins' speculative reconstruction.

Fig. 10.4 Hawkins' Crystal Palace *Labyrinthodon* appearing in other contexts. **A)** undated (possibly 1860s) watercolour sketch by Hawkins of an unnamed '*Labyrinthodon*' species alongside other Triassic animals; **B)** Owen's influential line art of '*L. pachygnathus*' from Lyell's 1855 Manual of Elementary Geology. Owen's 1840s work on labyrinthodonts and 1854 guidebook were non-committal on detailed aspects of their life appearance, but this sketch shows he was won over, to at least some extent, by Hawkins' Crystal Palace restoration.

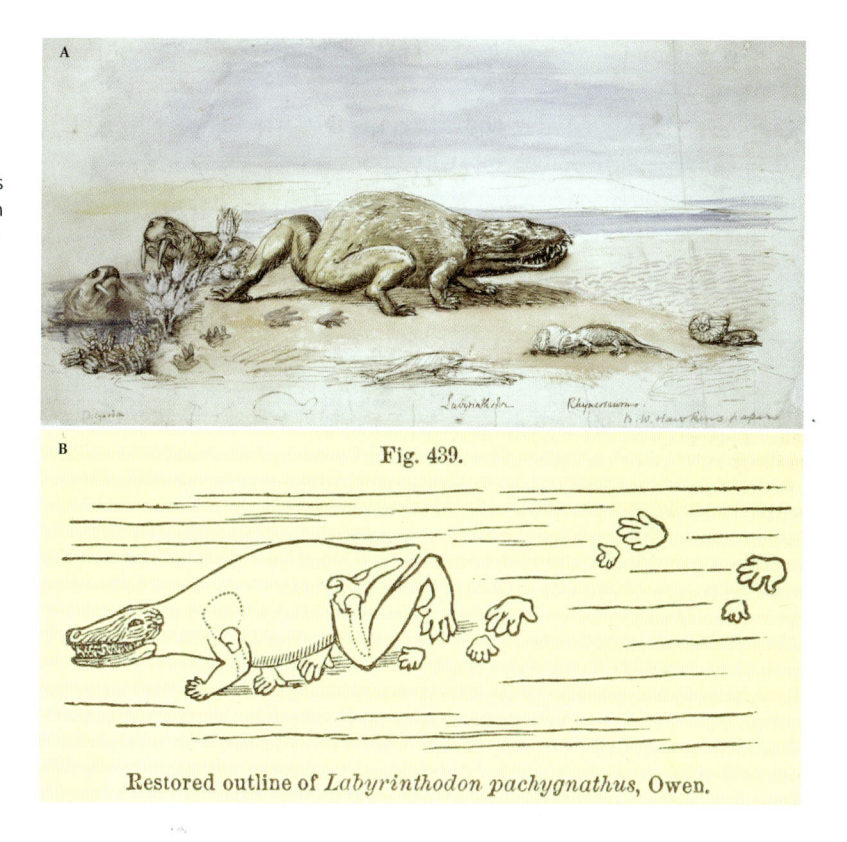

The body plans of the Crystal Palace '*Labyrinthodon*' species are similar. Although generally frog-like, Hawkins also incorporated reptilian elements in line with Owen's interpretation. Both are crouching animals with arching backs and large, wide heads. In '*L. pachygnathus*' only, a secondary row of teeth is present on the roof of the mouth. Their sharply arched backs presumably reflect the presence of a mobile vertebral junction between the pelvic and torso region, as per living frogs. Their legs are proportionally long and well-muscled, and terminate in elongate, well-jointed feet. The digits on both hands and feet are short, befitting amphibians which spend much of their time on land. A reptile-like feature which contrasts against both frogs and the reality of *Mastodonsaurus* are the short necks of both '*Labyrinthodon*' species, where all three sculptures have heads flexed relative to their bodies.

That the restorations of both species share many basic features is unsurprising given that they share reference fossils, particularly the *Chirotherium* tracks. Nineteenth-century restorations of this genus were rarely shown without these footprints (see Figs 10.3–10.4A) and this tradition extended to Crystal Palace: Hawkins cast *Chirotherium*-like tracks into the concrete anchoring at least one '*Labyrinthodon*' model in place (see Fig. 10.6). These were the only simulated footprints in the Geological Court and they have, sadly, vanished over time. The last one was still visible in 1994 (McCarthy and

Fig. 10.5 Details of the Crystal Palace '*Labyrinthodon*' in 2013 and 2021. Note the palatal teeth in '*L. pachygnathus*'.

Chapter 10 – The sculptures: '*Labyrinthodon*' 145

Gilbert, 1994) but this has now either weathered away or was filled during conservation works. A bizarre fact of Owen's amphibian *Chirotherium* hypothesis is that he rationalized the trackmaker as making prints with opposite limb sets, so the left prints were made with the right limbs and vice versa (he attempted to draw this in his own reconstruction, *see* Fig. 10.4B). This was demanded by the configuration of toes in the prints which were inverted from those seen in living amphibians. The strange gait required by this proposal is absent from Hawkins' models despite his general attention to Owen's other ideas: perhaps even Hawkins could not make this bizarre hypothesis a reality.

Owen was correct, however, in contemplating the disproportionate size of the *Chirotherium* footprints. Their trackmakers walked fully on their feet (i.e. they were plantigrade, placing their ankles on the floor to walk and stand) but only touched the ground with their fingers, such that their hand prints appear much smaller than their footprints. With little other information to work with, Owen's suggestion that frog-like creatures made such tracks was an inventive and sensible interpretation of the track size disparity, and one that would only be fully overturned in the 1920s when considerably more fossil data allowed the *Chirotherium* trackmaker issue to be readdressed (Bowden *et al.*, 2010). Since these studies it has been realized that *Chirotherium* was likely made by reptiles, not amphibians and, more specifically, that *Chirotherium* was probably left by Triassic pseudosuchians – the crocodylian-line of reptile evolution – that were adapted for walking fully upright like dinosaurs and mammals. This is not the only connection between pseudosuchians and Owen's '*Labyrinthodon*': we will find another shortly.

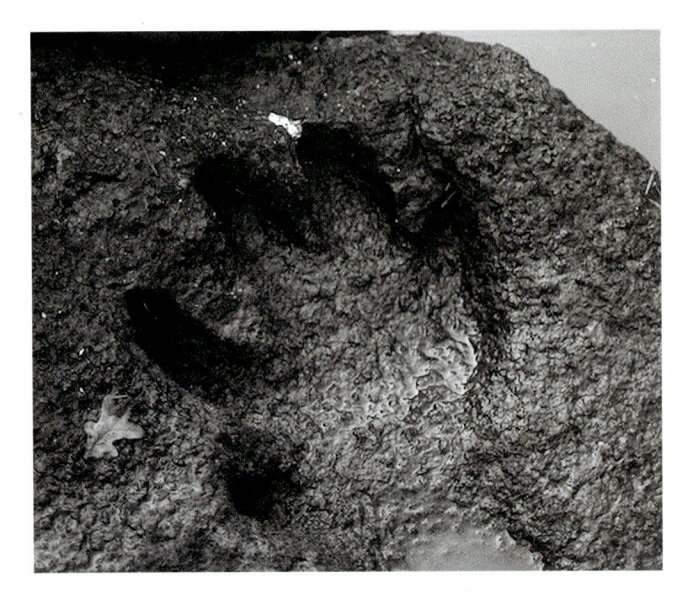

Fig. 10.6 One of the few photographic records of the *Chirotherium* tracks of the Geological Court, probably taken in the early 1990s. Since this image was captured, any remains of the tracks (including this example) have either weathered away or were covered during conservation work.

Fig. 10.7 The Crystal Palace 'amphibians' owe much of their anatomy to reptiles, specifically pseudosuchians – the crocodylian-line of reptile evolution. These mainly quadrupedal reptiles are the likely trackmakers of *Chirotherium barthi* and fragmentary bones of one type, the sail-backed ctenosauriscids, were factored into the reconstruction of '*L. pachygnathus*'. The ctenosauriscid in this reconstruction is *Arizonasaurus babbitti*, one of the best-known examples of this group.

'*Labyrinthodon salamandroides*'

The larger of the two '*Labyrinthodon*' species at Crystal Palace is '*L. salamandroides*', a 3.6m-long, smooth-skinned creature depicted as walking onto the shore of the Secondary Island (*see* Figs 10.1, 10.5). Owen (1854) outlined the fossil inventory behind '*L. salamandroides*' in detail, listing an entire skull, lower jaw, vertebrae and some limb fragments (*see* Fig. 10.2). Much of this collection pertained to German fossils of the large temnospondyl *Mastodonsaurus giganteus*, making this restoration a chimeric mix of this species and the pseudosuchian trackway *Chirotherium*.

Given that Hawkins had a complete skull and most of a lower jaw to work from, it is notable that his '*L. salamandroides*' head is an inexact match to the real cranial proportions of *Mastodonsaurus*. As with many temnospondyls, *Mastodonsaurus* had an extremely flat, broad and low skull with small openings for eyes and nostrils situated on the upper surface. Owen's (1842e) comparisons with crocodylians were apt, as these reptiles have similarly low, solidly built skulls. Hawkins has captured some of this in his model – the head is somewhat broad, flattens towards the jaw tip, and has some fossil-mimicking ornament on the lower jaw – but it is overall much too tall and has prominent, laterally-facing eyes and nostrils.

It's fair to regard this as one of the less successful restorations by Hawkins and we may wonder if he was overwriting fossil anatomy with that of living species – principally frogs – when designing this animal. We should consider, however, that achieving the very flat, broad skull shape of *Mastodonsaurus* may have required a more sophisticated sculpture with a metal head, as per the pterosaurs and mammals. While certainly achievable from an engineering perspective, perhaps the cost, time or another practical concern influenced the outcome of the model. In any case, this inexact reconstruction is probably more evidence of Owen's lack of guidance given to the project, despite his role in originating the 'sauroid batrachians' concept of '*Labyrinthodon*' and the availability of good fossil data.

The '*L. salamandroides*' sculpture is a longer-legged and more muscular variant of the '*Labyrinthodon*' body plan developed by Hawkins, and also possesses a more proportionate head. These features, along with the relatively narrow gait (referencing *Chirotherium*) and large appendages, suggest this model leans a little more towards the 'reptilian' end of the 'sauroid batrachian' brief and impresses an enhanced sense of terrestriality compared to '*L. pachygnathus*'. The positioning of the sculpture as emerging from the water implies that some degree of aquatic life was still imagined, although the exact ecology is not obvious: the reconstruction has qualities recalling frogs, crocodylians and even bears.

Fig. 10.8 A modern take on *Mastodonsaurus giganteus*, a fossil species strongly informing '*Labyrinthodon salamandroides*'. These 5–6m-long animals were some of the largest 'amphibians' of all time and operated in crocodylian-like ecologies.

Why Hawkins elected to depict '*L. salamandroides*' with more terrestrial characteristics is not clear. Perhaps he simply wanted to differentiate the model from '*L. pachygnathus*' in aspects other than size? Whatever the reason, the *Mastodonsaurus* bones informing his restoration actually belonged to a far different creature: a wholly aquatic animal resembling a salamander-crocodile cross with a very large, flat head, several rows of teeth, large fangs, a powerful sculling tail and short limbs. Hawkins was correct to interpret '*L. salamandroides*' as a large animal, however, as *Mastodonsaurus* was among the largest amphibians to have ever existed.

'*Labyrinthodon pachygnathus*'

Two smaller '*Labyrinthodon*' species, the 'thick jawed' *pachygnathus*, are positioned alongside '*L. salamandroides*', one facing out from the water margin, one facing inland. These 2.6m-long sculptures have noticeably larger heads and smaller limbs than their larger neighbour, as well as differently shaped, flatter and deeper jaws. A rugose, knobbled texture extends across much of the body, including the head, to create a gnarled, weathered appearance. Owen's (1854) guidebook offers only a brief description of this species but one of the guidebook illustrations (*see* Fig. 10.2) shows that the fossils informing the reconstruction match those described in Owen's earlier (1842e) descriptive work on '*L. pachygnathus*', including broken skull and jaw bones, teeth, vertebrae, a partial pelvis and fragments of limb bones. These distinctive fossils demonstrate that the '*Labyrinthodon*' figure in *Geology and Inhabitants of the Ancient World* must be mislabelled: although captioned as '*salamandroides*', it clearly has the fossil inventory of '*pachygnathus*'.

'*L. pachygnathus*' is a more complex chimera than '*L. salamandroides*'. The skull and jaw remains represent generically indeterminate mastodonsaurid material (Damiani, 2001) but the other fossils represent pieces of reptile. Today, we recognize these as belonging to *Bromsgrovia walkeri*, a type of pseudosuchian belonging to the ctenosauriscid group (Benton and Gower, 1997). Ctenosauriscids were remarkable reptiles that resembled tall, upright crocodylians and bore large sails along their backs (*see* Fig 10.7). Somewhat fittingly, they are part of the Rauisuchia, a major radiation of pseudosuchians that were probably responsible for creating *Chirotherium* tracks (Bowden *et al.*, 2010). Thus, although outwardly frog-like in appearance, the body and trackway fossils informing Hawkins '*L. pachygnathus*' were primarily reptilian. It may, accordingly, be more apt to regard the two '*pachygnathus*' sculptures as early efforts to restore pseudosuchians rather than prehistoric amphibians.

As with '*L. salamandroides*', Hawkins seems to have taken heavy liberties with his reconstruction. The head of '*L. pachygnathus*' is also not especially mastodonsaurid-like, with its deep snout and laterally-situated eyes. The distinctive skin of the '*L. pachygnathus*' statues seems, at first glance, to approximate the warty skin of toads and other leathery skinned frogs, and further frog-like features are the large tympanum (ear openings) at the rear of the heads and horizontal pupils. The knobbled skin, however, may reflect another influence: Owen's prediction that '*L. pachygnathus*' was covered in bony plates, in line with the crocodile-like sculpting of its jaw bones:

> No anatomist, indeed, can contemplate the extensive development and bold sculpturing of the dermal surface of the cranial bones in the *Labyrinthodon pachygnathus* ... without a suspicion that the same character may have been manifested in bony plates of the skin in other parts of the body.
>
> OWEN, 1842E, P. 541

The true nature of temnospondyl skin remains unevidenced, but it is thought that their skull textures may reflect once being covered with tight, crocodylian-like skin (Witzman, 2009). Although they were not generally armoured, Owen's prediction of bony structures in the skin of some prehistoric 'amphibians' was not wrong: the mastodonsaurid *Sclerothorax* had numerous tiny ossicles embedded in its skin and one group, the dissophoroids, sported large bony plates on their backs. The Carboniferous and Permian dissophorids were squat, strongly terrestrial creatures best known from *Caecops* and the sail-backed *Platyhystrix*, species not anatomically dissimilar to the 'sauroid batrachians' imagined by Owen and restored by Hawkins. As with Owen's dinosaurs, hindsight shows that the science behind Owen's '*Labyrinthodon*' was flawed but the Crystal Palace reconstructions excellently demonstrate nineteenth-century open-mindedness about life of the past.

CHAPTER 11

The sculptures: *Dicynodon*

The last of the Geological Court's palaeontological sculptures are situated on the southeastern extent of the Secondary Island. Although still relatively large statues, both of these squat reconstructions face away from paths around the Island and are easily overlooked and forgotten by park visitors focused on the more spectacular prehistoric residents of Crystal Palace. They represent two species of *Dicynodon*, a genus named by Owen in 1845 for tusked skull remains blending mammalian and superficially 'reptilian' features.

Found in the now famous Karoo Supergroup of South Africa, *Dicynodon* was the first fossil animal recognized as a 'stem-mammal': the diverse and abundant, somewhat reptile-like creatures that ultimately gave rise to our own mammal lineage. Unbeknownst to nineteenth-century palaeontologists, other stem-mammals (dinocephalians) had already been discovered in Russia, but their mammalian qualities had not yet been identified. *Dicynodon* was, therefore, the animal which delivered the breakthrough insight that 'reptile-grade' animals had something to do with the ancestry of mammals (Kammerer *et al.*, 2011). Initially assumed to have come from Triassic rocks, revisions to our understanding of the Karoo deposits show that *Dicynodon* was actually a Permian taxon. It is thus aptly placed at the very end of the Geological Court sculpture sequence as the geologically oldest genus in the display.

The *Dicynodon* are among the most difficult sculptures to see on the Geological Court, in part because of their size and low stature, as well as their orientation away from visitors, depicted as if they are crawling out of the waters surrounding the Secondary Island. They have survived to the modern day despite a moderate amount of damage to their principally concrete and brick frames, including major cracks and fissures in their bodies and the destruction of some limbs (since restored). As is typical with the water-adjacent sculptures, their original colour cannot be deduced due to rapid deterioration of the original paint and, as inconspicuous statues, they have not been routinely photographed. Our knowledge of their historic colouration is thus poor, but in recent years they have been painted with grey carapaces and grey-green bodies.

The identity of the *Dicynodon* species featured at Crystal Palace was not made explicit during the Crystal Palace era. Two species are present, this being obvious from their size difference as well as anatomical details that are best appreciated at close inspection, but only the identity of the larger model was revealed by Owen's guidebook. This 3.6m-long statue – 'with the bulk of a walrus' (Owen, 1854, p. 39) – is *Dicynodon lacerticeps*, but the identity of the smaller, 2.6m-long animal is unmentioned. This represents a potential problem for identification as, by 1854, Owen had already

Fig. 11.1 The terminal palaeontological sculptures of the Secondary Island and by far the most speculative sculptures of the entire display: *Dicynodon lacerticeps* (right) and '*Dicynodon strigiceps*' (left). (2015)

Fig. 11.2 Details of the *Dicynodon* models, from 2013 and 2021. Scientifically speaking, the most successful parts of the sculptures are their faces which are, save for the over-long snouts, fairly good approximations of dicynodont crania. Their bodies and limbs, however, are examples of unsuccessful anatomical correlation.

named three species of *Dicynodon* and more were on their way for an 1855 publication: the list of possible identifications for this sculpture was already long when it was completed. Indeed, like many other genera represented at the Crystal Palace, *Dicynodon* was starting a journey to becoming a wastebasket taxon: in this case, a particularly notorious one comprising 168(!) species. This unwieldy number flummoxed and frustrated decades of dicynodont research until the genus was thoroughly reviewed by Christian Kammerer and colleagues in 2011.

Thankfully, Owen's guide provides a clue that allows us to confidently identify the smaller sculpture: he states that it represents a species with owl-like facial features. This is surely '*Dicynodon strigiceps*', a species Owen named in 1845 with a name that translates to 'owl-faced *Dicynodon*'. Resolving this issue doesn't remove all confusion around the identities of these sculptures, however, as their identifications are peculiar in light of what was known about *Dicynodon* in 1854. *D. lacerticeps* was not regarded as a walrus-sized species but Owen (1845) applied this exact size descriptor to the large taxa '*D*.' *bainii* and '*D. tigericeps*' (now both *Aulacephalodon bainii*, Kammerer et al., 2011). Of relevance here is that the *Dicynodon* text in *Geology and Inhabitants of the Ancient World* is the final, shortest and least informative section in the book. Even more so than the rest of the guide, it has a sense of being rushed and careless – not only, for instance, omitting to name '*D. strigiceps*' but also erroneously suggesting that this sculpture was only partly restored – and it would not be surprising if it contains fundamental errors in reporting.

From a palaeoartistic perspective, the boldness of the Crystal Palace *Dicynodon* restoration is striking. To anyone who knows anything about dicynodonts – which are heavy-set, quadrupedal terrestrial herbivores in reality – the turtle-like appearance of Hawkins' versions is an immediate

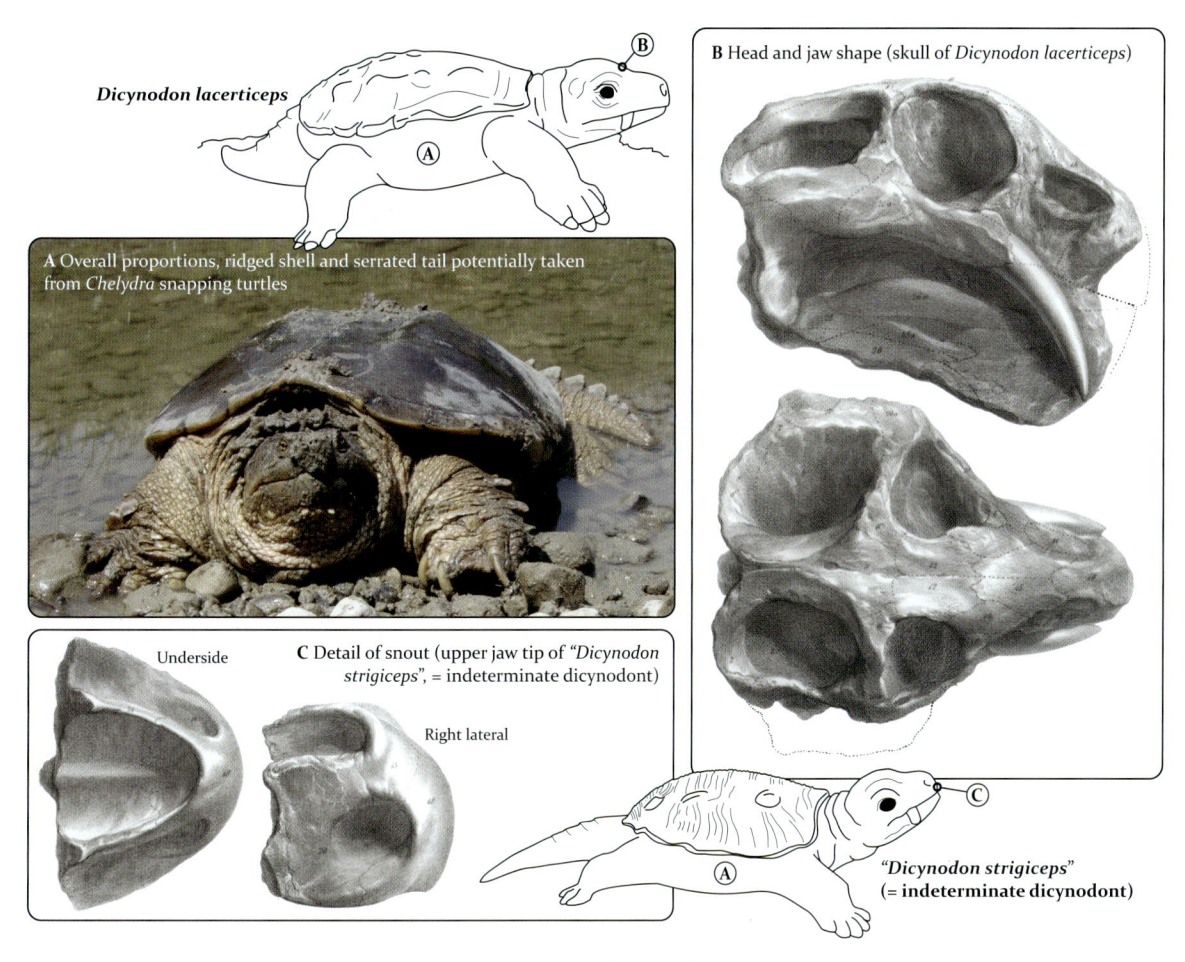

Fig. 11.3 The sparse fossil material and living animal species likely referenced for the Crystal Palace *Dicynodon*. Skull images from Owen (1845).

indication that they are the most speculated reconstructions of the entire Geological Court. The fact that Hawkins and Owen were confronted with an entirely new type of animal, represented by a minimal amount of fossil data, did not prevent their efforts at reconstruction, however, even though it resulted in a sculpture that was essentially guesswork. Hawkins' *Dicynodon* may therefore not only have the accolade of 'most speculated reconstruction' in the Geological Court, but also rank among the most conjectural reconstructions of this already quite experimental era in palaeoart history.

The entire fossil inventory factored into these restorations was two partial skulls (with an emphasis on 'partial' for '*D. strigiceps*') and a handful of vertebrae (Owen, 1845, 1854). This collection makes a mockery of the suggestion that the Crystal Palace sculptures were based on animals 'for which the entire, or nearly entire skeleton had been exhumed in a fossil state' (Owen, 1854, p. 4) and the addition of such a poorly known animal to the display is actually rather curious: were Hawkins and Owen so convinced of the power of anatomical correlation that they assumed these mysterious animals could be restored from skulls alone? Or were these sculptures the product of more practical concerns, such as fleshing out the southern region of the Secondary Island and portraying a greater sense of life's history and development? So far as scholars of the mid-nineteenth century knew, few large land animals were present in the New Red Sandstone, so options for animal sculptures for this region would have been limited.

The heads of Hawkins' *Dicynodon* are closely based on their skulls such that, even today, they are reasonable takes on this part of dicynodont anatomy (*see* Fig. 11.3). Their snouts are slightly long relative to their fossils – likely a consequence of poor preservation of the first discovered *Dicynodon* skulls – but faithfully restore their broad beaks and peculiar tusks. Behind the heads, however, is pure palaeoartistic speculation. The blend of turtle and walrus cranial features and the shape of *Dicynodon* vertebrae suggested to Owen that these animals were amphibious (Owen, 1845, 1854), and it was this that led to the turtle-like anatomy

Chapter 11 – The sculptures: *Dicynodon* **151**

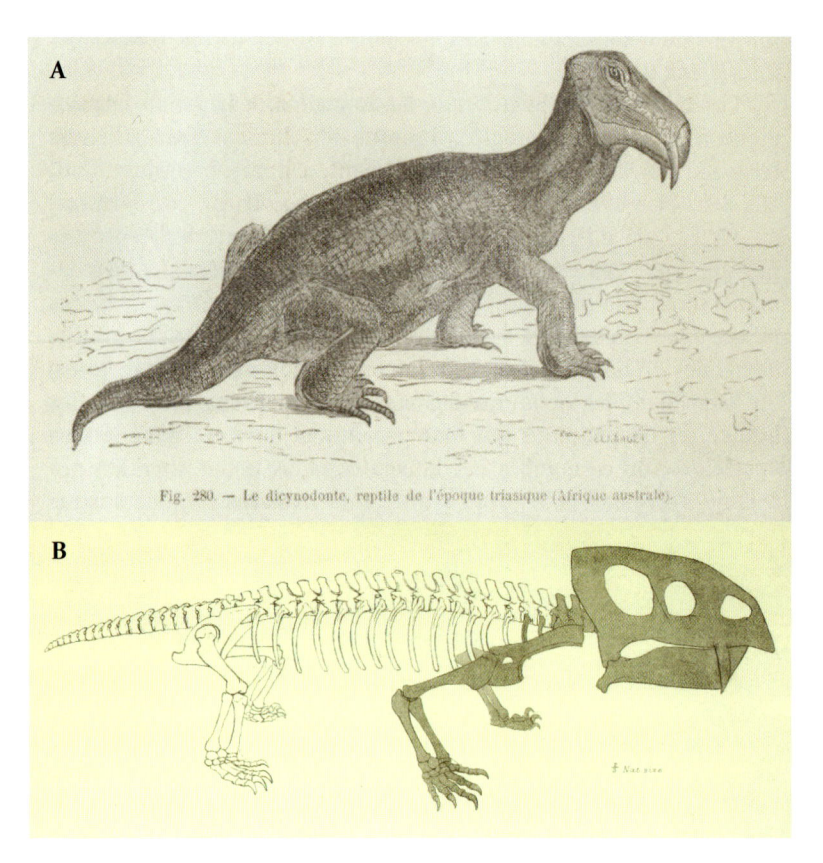

Fig. 11.4 The evolution of *Dicynodon* in palaeoart, post-Crystal Palace. Hawkins' speculative takes on *Dicynodon* never caught on, and it wasn't until more substantial material of dicynodonts became available that further efforts were made to reconstruct them. **A)** dicynodont from Flammarion's (1886) *Le Monde Avant la Création d'Homme*; **B)** composite skeletal reconstruction by Harry Seeley (1888) based on material from South Africa and Russia. Seeley was unhappy with several aspects of this work, but it was a significant step forward from the origins of dicynodont palaeoart.

for their bodies being restored 'conjecturally, to illustrate the strange combination of characters manifested in the head' (Owen, 1854, p. 39). Details of the sculptures' feet show that Hawkins was probably not modelling these creatures on fully aquatic turtles, an idea also implied by the keels, bosses and ridges adorning their shells and their long, serrated tails. These details were probably taken from extant turtles, with the common snapping turtle *Chelydra serpentina* potentially being a major inspiration (*see* Fig. 11.3).

These turtles also share the slightly undersized-looking shell with Hawkins' *Dicynodon*, and their diet, which is omnivorous but includes powerful acts of predation, would have chimed with Owen's interpretation of *Dicynodon* as a carnivorous animal (his explanation for their mighty tusks). The *Dicynodon* are not entirely turtle-like, however, as they have a seemingly solid, one-piece shell rather than one made from a complex of interlocking scutes, and they also lack a plastron (the shell covering the underside of the body). Perhaps Hawkins did not want to make his restorations wholly turtle-like and aimed only to borrow enough testudine anatomy to convince visitors that they are looking at real animals and not – as was the case – mostly imaginary ones.

Hawkins' *Dicynodon* stood as some of the few reconstructions of dicynodonts for much of the nineteenth century, although sufficient material was seen by Owen that, by 1862, he was able to predict a low-slung animal with a heavy body. Owen never attempted to rationalize *Dicynodon* life appearance beyond this, however, and it was not until the late 1880s that reconstructions based on fossils from Russia and South Africa saw dicynodont anatomy approximated in art for the first time (*see* Fig. 11.4; Flammarion, 1886; Seeley, 1888). Subsequent reconstructions, most notably by Helga Pearson (1924), refined the visage of these animals as we entered the twentieth century. These works focused on different species of dicynodonts, however, leaving the life appearance of *Dicynodon* itself something of a mystery until the recovery of a near-complete *D. lacerticeps* skeleton in the 1970s. After more than a century, this specimen finally provided us with a general understanding of its shape and form (*see* Fig. 11.5; King, 1981).

Such complete *Dicynodon* specimens complement what is now a very developed understanding of dicynodont anatomy and diversity. They also show how wide of the mark Hawkins' sculptures were. While their heads are reasonable proxies for dicynodont crania, the widely-sprawled, tortoise-like limbs and bodies contrast with the long, narrow torsos and semi-sprawled, terrestrially adapted limbs of *Dicynodon* and its kin. Insights into dicynodont skin remain elusive but we can be certain that they lacked turtle-like bony shells and armour. Given their relationship to mammals, some modern

Fig. 11.5 A modern take on the life appearance of *Dicynodon lacerticeps*: a mid-sized creature with the same tusked, beaked jaw depicted by Hawkins, but a distinctly different, somewhat pig-like body.

artists have taken to restoring dicynodonts with fur, but we have yet to find any evidence of this integument type so deep within our evolutionary history, and the only scraps of skin we have from even remotely closely-related Permian stem-mammals are hairless (Chudinov, 1968). The recovery of hair from a Permian coprolite (Bajdek *et al.*, 2016) suggests some mammal-ancestors from this time may have been furry, but the most likely owners of this hair were our closest Permian ancestors, the cynodonts, not the more distantly related dicynodonts.

Dicynodon lacerticeps

The larger *D. lacerticeps* model represents the first named species of *Dicynodon*, *D. lacerticeps*. This species is one of only two valid *Dicynodon* species recognized today and is represented by a large number of fossils that include good skulls, limb material and even near-complete skeletons (Kammerer *et al.*, 2011). At 3.6m-long, the model is considerably bigger than the neighbouring '*D. strigiceps*' and, as noted above, the rationale behind this decision is peculiar. Large skulls now referred to the genus *Aulacephalodon* were probably factored into the build. Actual *D. lacerticeps* specimens suggest an animal around 1m long (King, 1984), such that the smaller '*D. strigiceps*' model is a better size match.

Other than size and a carapace texture, there is little to distinguish the Geological Court *Dicynodon* from one another: even their poses are nearly identical. Although positioned close together, neither animal is dynamically arranged or placed in a manner which implies movement or intent, as is common around the rest of the display. Presumably the desired effect was to show these animals basking, like turtles, but the impression it leaves visitors with is slightly underwhelming in light of the other spectacular arrangements in the Geological Court.

'*Dicynodon strigiceps*'

The smaller of the two *Dicynodon* models represents a very poorly known species suggested by Owen to have an owl-like face – a somewhat questionable conclusion given that only the snout tip of this species was known to him. Little evidence of owl-like features are seen in the face of Hawkins' model, which measures 2.6m long. '*D. strigiceps*' has long been regarded as an invalid species by dicynodont workers as the remains are too poor to show even basic distinguishing features: it cannot even be classified to a particular type of dicynodont, let alone a specific genus or species (Kammerer *et al.*, 2011). Contrary to the Crystal Palace model, the original specimen appears to lack tusks, which tentatively recalls the tuskless dicynodon *Oudenodon bainii* (Kammerer *et al.*, 2011). Such an identification cannot be confirmed, however, given the scant nature of the material this species was founded upon.

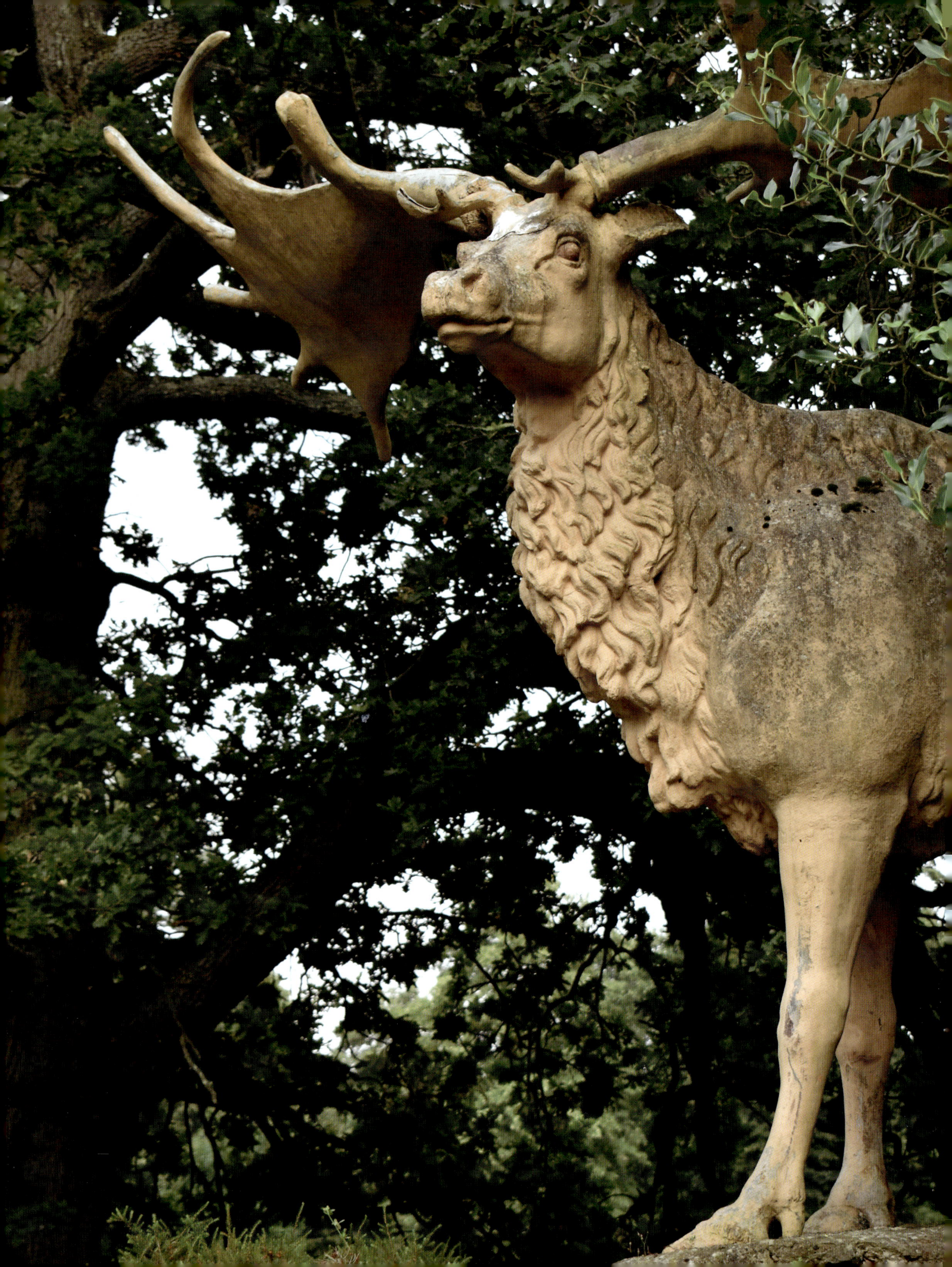

PART 3

A difficult...and, perhaps, too bold, attempt

In every stage of this difficult, and by some it may be thought, perhaps, too bold, attempt to reproduce and present to human gaze and contemplation the forms of animal life that have successively flourished during former geological phases of time, and have passed away long ages prior to the creation of man, the writer of the following brief notice of the nature and affinities of the animals so restored feels it a duty, as it is a high gratification to him, to testify to the intelligence, zeal, and peculiar artistic skill by which his ideas and suggestions have been realised and carried out by the talented director of the fossil department, Mr. Waterhouse Hawkins. Without the combination of science, art, and manual skill, happily combined in that gentleman, the present department of the Instructive Illustrations at the Crystal Palace could not have been realised.

RICHARD OWEN, 1854, P. 7

Having looked at the creation, construction and scientific approaches of the Geological Court, our final section turns to what happened next: the 170 years that have passed since the Crystal Palace opened its gates to the public. It perhaps goes without saying that the fate of the Geological Court has been inextricably tethered to the eventful and often tumultuous history of the wider Crystal Palace project. The financial challenges of the Crystal Palace Company, the changing nature of the park landscape, and several transfers of land ownership loom large in this stage of our narrative. Against this backdrop of local challenges and upheaval, the impact of the Crystal Palace Dinosaurs was felt on the international stage, imbuing them with varying amounts of popular, media and academic interest that ranged from praise and laudation to harsh criticism and ridicule.

Both the park history and reception to the site have influenced approaches to what is now surely the most critical aspect of the Geological Court: its long-term survival against a myriad of conservation risks. These have already taken their toll on several aspects of the display and the site as we know it today is not that originally built by Ansted, Campbell, Hawkins and Owen. Rather, it is a modified, reconstructed and incomplete version that has persisted to modern times because of periodic intervention against destructive natural and human agents. Any future longevity will rely on similar, and perhaps even more drastic restorative action, such that the continued existence of the Geological Court should not be taken for granted.

CHAPTER 12

The reception and legacy of the Geological Court

In June 1854 the Geological Court was opened to the public along with the rest of the Crystal Palace. Hawkins, Ansted and Campbell were still only half-finished constructing the geological landscape and models when the public first saw their work but they drew a strong and immediate reaction nonetheless. That the Crystal Palace Dinosaurs had a substantial impact on palaeontological culture is well known, but the full complexity of their legacy is not easily summarized. The greatest fans of the Geological Court were initially laypeople and the press, but these devotees became increasingly indifferent and uninterested in the following decades. Conversely, some academics and educators were fierce critics of the Geological Court when it first opened, but their views mellowed to more considered and even praiseworthy commentary in later decades. The models were both influential on nineteenth-century palaeoart in that their designs were copied and reproduced in numerous illustrations, but their rapid obsolescence may have also made scholars nervous about reconstructing newly discovered taxa. If we must boil the Geological Court project down to some arbitrary measure of success, they have never been regarded as a total triumph nor total failure.

The Geological Court opens: what happened next

The debut of the new Crystal Palace in June 1854 was, by all accounts, tremendously well received and accompanied by much media fanfare and ceremony. Among all the praise for the rebuilt Palace and exhibitions, the Geological Court proved to be a frequent topic of press coverage and a popular public attraction even as Hawkins, Ansted and Campbell raced to finish the models and grounds. This heightened interest must have been a relief to the Crystal Palace Company given the novelty, ambition and expense of the Geological Court. Some 40,000 people were said to attend the opening of the Palace grounds and such a strong chord was struck with park visitors that Hawkins, despite his relatively low rank in the Crystal Palace Company, was among the select few asked to guide royal visitors around the park during opening events (McCarthy and Gilbert, 1994).

Media attention for the Geological Court, already high during its production, continued once the park opened. Many articles were written about the recreated extinct animals and, inspired by the work of Hawkins, Ansted and their teams, also covered topics such as Deep Time and the palaeobiology of the species displayed in the park. The fantastic appearance of the sculptures even saw them featured in cartoons, most famously in the satirical and humour

Fig. 12.1 Public reaction to the Crystal Palace Dinosaurs was tremendous despite many park guests not fully grasping the meaning of the unusual animals and geological arrangements. They became so popular that they entered Victorian pop culture, even appearing in the famous satirical magazine *Punch* on several occasions (*Punch* pieces shown here from Volume 28, 1855).

THE EFFECTS OF A HEARTY DINNER AFTER VISITING THE ANTEDILUVIAN DEPARTMENT AT THE CRYSTAL PALACE.

A VISIT TO THE ANTEDILUVIAN REPTILES AT SYDENHAM—MASTER TOM STRONGLY OBJECTS TO HAVING HIS MIND IMPROVED.

magazine *Punch* (Rudwick, 1992). This media interest has been interpreted by some as the Crystal Palace Dinosaurs being the major catalyst in the public's fascination with dinosaurs, often termed 'dinomania' (e.g. Torrens, 2012), but three points argue for a more measured perspective.

Firstly, although dinosaurs were undeniably a popular component of the display, they were not always the focus of media attention. Coverage was often evenly split among the extinct species and, surprisingly, the pterosaurs were often among the most extensively discussed if, conversely, rarely illustrated. Secondly, culturomics – the study of social trends through quantitative analyses of texts – show that prehistoric animals, including the dinosaurs *Iguanodon* and *Megalosaurus*, were already widely featured in books and periodicals before 1854 (*see* Fig. 1.9; Nieuwland, 2019), and a variety of richly illustrated texts devoted to prehistoric life were already on sale to the public (Rudwick, 1992). There was clearly already an existing appetite for palaeontological content among the public before the Geological Court was built. Thirdly, although some international interest was

Fig. 12.2 The Geological Court can be seen as the birthplace of modern palaeontological merchandise. Among the most famous examples of tie-in products are 1:12 scale models of select extinct species made by Hawkins in the 1850s, which were sold by James Tennant as educational aids. Bronze replicas of four of the five models are shown here: **A)** *'Labyrinthodon'*, **B)** *Temnodontosaurus*, *Plesiosaurus* and *'Plesiosaurus' macrocephalus*; **C)** *Iguanodon*; **D)** *Megalosaurus*. The fifth model (not shown) represents a pterosaur, of which only a few examples survive in museums.

shown in the Geological Court, most media focus was contained within the UK. Culturomic trends show that dinosaurs became popular in nations such as France and the United States only when their fossils were discovered closer to home, suggesting a nationalistic property to their celebrity (Nieuwland, 2019). Rather, international 'dinomania' as we know it today was truly achieved much later in the nineteenth century through canny marketing led by American scientific and entertainment institutions (Secord, 2004). The Crystal Palace Dinosaurs undeniably escalated public and media interest in dinosaurs and prehistory in the UK, but they were a prelude to the dinosaurian dominance of popularized palaeontology, not the start.

This is not to say that the Geological Court did not have a lasting impact on the public presentation of prehistory. Indulging in hobbies such as geology and palaeontology, or purchasing the expensive books on such topics, was the reserve of the rich in the mid-1800s. By making these topics accessible to the less wealthy and educated, the Geological Court had a democratizing effect for these subjects (Lescaze, 2017) and foreshadowed the cultural enthusiasm dinosaurs would eventually achieve in later decades. Hawkins' models also refined the portrayal of prehistory, combining hitherto unreached palaeoart spectacle with, almost paradoxically, presentation of a more realistic, grounded view of extinct animals than the fantastical artworks that preceded them (*see* Fig. 1.9). These were extinct creatures that looked like real animals, regardless of their scientific precision, and their relatability surely aided their reception (Nieuwland, 2019).

Even before the Geological Court was opened the merchandising potential of the Crystal Palace Dinosaurs was being discussed. On 17 May 1854, just weeks before the Crystal Palace opened to the public, Hawkins delivered a lecture to the London Society of Arts to promote the Geological Court and explain its construction. The transcript of this event not only preserves some of the first impressions of Hawkins' work (*see* below) but also the enthusiasm for having his models transferred into smaller, affordable formats for educational use. This discussion, based around bringing Hawkins' work to schools and museums, as well as toy shops, has been seen as 'ground zero' for the many extinct animal models and collectables that now fill museum gift shops and toy stores (Rudwick, 1992; Liston, 2010).

The first recognition of this educational business opportunity was Richard Dawesby, Dean of Hereford, who suggested that smaller versions of the sculptures could be sold for 'a moderate price' to those unable to travel to see Hawkins' models for themselves (Hawkins, 1854, p. 445). This idea was enthusiastically received among other attendees of Hawkins' lecture, including Professor James Tennant, a Professor MacDonald and Mr Harry Chester. When discussing the many formats that Hawkins' sculptures could be translated into, Chester made one cautioning comment which suggests the renowned steely resolve of Victorian gentlemen may have become somewhat exaggerated with time:

But there was one form of illustration which he hoped this subject would not receive, but which he feared would be the case, that was, that these monsters would find their way into their carpets and paper hangings. He would ask what would be the consequence, if a gentleman of not very strong nerves, on plunging into his bath found the bottom of it ornamented with some of these horrid-looking animals.

CHESTER (QUOTED IN HAWKINS, 1854, P. 448)

Chester could bathe easy, however, for no bathtubs seem to have been adorned with Hawkins' extinct creatures. It was instead agreed that 1:12 scale versions of the giant models would be sculpted by Hawkins and sold by audience member James Tennant, who was also a noted geologist and purveyor of geological specimens (McCarthy and Gilbert, 1994). Plaster models that represented the *Iguanodon*, *Megalosaurus*, Jurassic marine reptiles, a pterosaur and '*Labyrinthodon*' were the result, and examples of these models (or replicas) survive today in museums around the world. Later, the American businessman Henry Ward sold a similar series of models in his famous 1866 *Catalogue of Casts of Fossils* (Davidson, 2005). The relationship of Ward's models to the Tennant/Hawkins versions is uncertain. They are sometimes regarded as direct casts but Davidson (2005) was unable to find any clear evidence of this, and suspects they were derivative works closely based on Hawkins' originals. Nevertheless, the range of models and their catalogue illustrations are essentially identical to the models available in the UK so, if they were copies, they were very close approximations.

Fig. 12.3 Pages from the 1866 *Catalogue of Casts of Fossils* detailing Ward's Crystal Palace-based extinct animal models, which were either cast from or copied directly from the 1:12 scale models created by Hawkins.

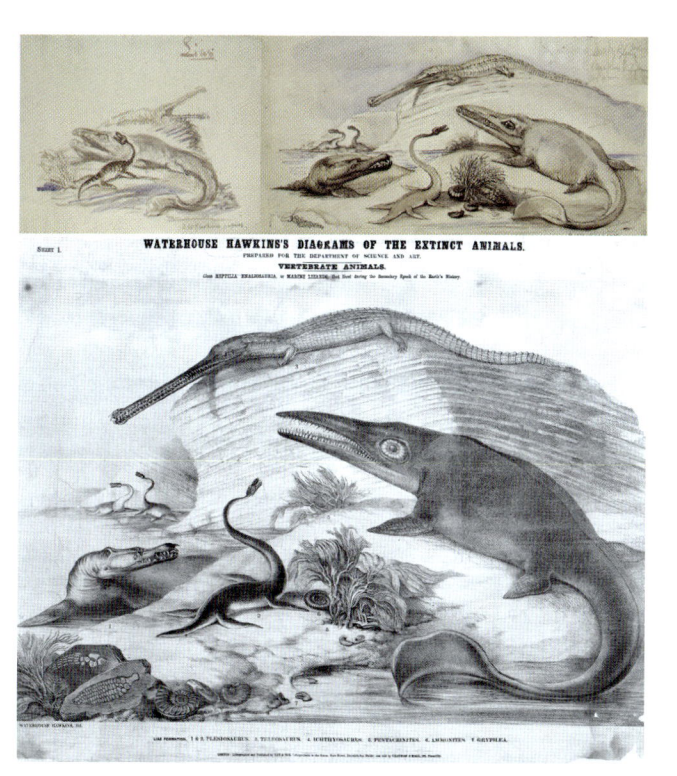

Fig. 12.4 The Geological Court in paper form: Sheet 1 of Hawkins' c. 1862 poster series of extinct life, the first five of which are heavily based on his Crystal Palace work. Archived illustrations show the plans for this layout of enaliosaurs and *Teleosaurus*, which appears to be perched on a Lias cliff. Note the large-headed '*Plesiosaurus macrocephalus*' and *Teleosaurus* scale pattern, which is still uncorrected from that at Crystal Palace.

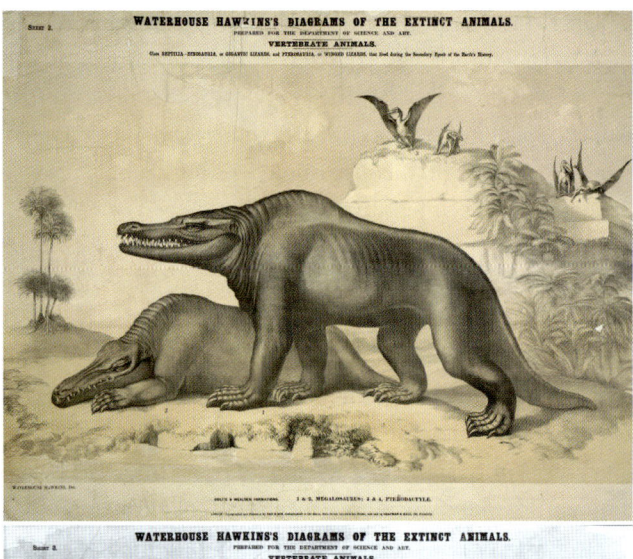

Fig. 12.5 Sheets 2 and 3 of Hawkins' c. 1862 poster series, showing *Megalosaurus* and pterosaurs in front of limestone cliffs, and *Iguanodon* and *Hylaeosaurus*, respectively. The compositional similarities to the Secondary Island are obvious.

Fig. 12.6 Sheets 4 and 5 of Hawkins' c. 1862 poster series, showing *Anoplotherium*, *Xiphodon* and *Palaeotherium*, and *Megatherium* and *Glyptodon*, respectively.

Further Geological Court-related products were created by Hawkins in the early 1860s: six posters featuring his Crystal Palace species arranged around appropriate landscapes (see Figs 12.4–12.6). These were also sold by Tennant for educational purposes, and the posters were augmented with species not featured at the Geological Court, including a whole poster devoted to mammoths and other Pleistocene mammals (Rudwick, 1992).

The cancellation of the Geological Court project

With public interest in the palaeontological models so high, it must have seemed unquestionable that Hawkins and Ansted would be given the additional funds and resources needed to realize the entire vision of the Geological Court: principally, a finished Tertiary Island fully populated with prehistoric animals and fully developed geological features.

A rough map drawn by Hawkins in 1855 reveals the extent of the original plan (see Fig. 12.7), and press articles made these ideas known to the public. They detailed numerous new sculptures including snakes and turtles (including '*Colossochelys*', today known as *Megalochelys*), the giraffe relative *Sivatherium*, the chin-tusked elephant-relative *Deinotherium*, several species of bovid, lions and tigers, *Glyptodon* (a large, armadillo-like mammal), woolly mammoth, mastodon, dodo, and three moa. Alongside these, new geological formations were mapped out. The Secondary Island would have additional Mesozoic deposits (Upper and Lower Greensand, and Gault Clay), with a small area of Chalk shown on the Tertiary Island alongside new Cenozoic formations: the London Clay, Paris Basin, Calciere Grossier (= Lutetian limestone), and a series of 'Lower Tertiaries' (a collection of rocks in the southeast UK). This radical plan

160   Chapter 12 – The reception and legacy of the Geological Court

Fig. 12.7 Sketch map sent by Hawkins in 1855 to Owen as part of a letter about the cancellation of the Geological Court project. This is our most detailed record of the planned layout of the unfinished Tertiary Island.

would have transformed the Tertiary Island and made the mammal-dominated section of the Geological Court a rival to the dinosaurs in not only spectacle, but also educational value.

Unfortunately, this expansion never happened. By September 1855 the Crystal Palace Company was finally feeling the financial impact of their epic undertaking and, despite the rudimentary working conditions for the Geological Court build team and Hawkins being paid 'little more than journeyman's wages' for his efforts (Dawson, 2016), they could not afford the expenses of the remaining court content. The total cost to the Crystal Palace Company for constructing the Geological Court and its prehistoric animals was already substantial at £13,792 (McCarthy and Gilbert, 1994; equivalent to £1.5 million today) and, although just a fraction of the total debt accrued by constructing the entire Crystal Palace site, developing the Tertiary Island was simply too much additional capital for company accountants. At this point all but the most heavily invested or essential projects in the park and gardens were being cancelled and the planned Tertiary Island expansion, with well over sixteen new sculptures and at least seven Geological Illustrations, was beyond what the floundering Crystal Palace Company could afford.

The abruptness with which development on the Geological Court ended seems shocking. Hawkins' contract was terminated so sharply that he was not even permitted to finish the half-completed mammoth model, which already had foundations in the park and might have been completed for just a few hundred pounds (Dawson, 2016). This decision was criticized by prestigious friends of the Crystal Palace as well as the public. In September 1855 *The Observer*, a strong proponent of the Geological Court, expressed their dismay at the cessation of work on the Geological Court and lamented the loss of the mammoth:

> The mammoth is doomed to obscurity; the huge framework of the mighty beast which now stands in the shed upon the island is to be resolved into its constituent elements; and of that colossal elephant... there is to be no representative... Considerable sums have already disbursed in the partial construction of the mammoth, and, as but a small additional amount would be required for its completion, it certainly appears to be a measure of unwise economy thus to throw away the sums which have been expended.
>
> THE OBSERVER, 17 SEPTEMBER 1855, P. 5

Fig. 12.8 Select oil paintings from Hawkins' tenure at the College of New Jersey, painted 1876–77. This series of seventeen images was Hawkins' last major palaeoart project and saw him returning to many species he first recreated at Crystal Palace. **A)** *Triassic Life of Germany*, with *'Labyrinthodon'* and *Nothosaurus*; **B)** *Early Jurassic Marine Reptiles*, with *Temnodontosaurus* and *Plesiosaurus* (1876); **C)** *Jurassic Life of Europe*, with *Megalosaurus, Iguanodon, Teleosaurus*, pterosaurs and *Hylaeosaurus*.

Fig. 12.9 More 1876–77 oil paintings from Hawkins' College of New Jersey project, featuring species included at Crystal Palace and also several planned Geological Court taxa. **A)** *Cretaceous Life of New Jersey* (1877), with mosasaurs (probably *Tylosaurus*) and pterosaurs, along with *'Laelaps'* and *Hadrosaurus*; **B)** *Tertiary Mammals of Europe*, with *Anoplotherium, Palaeotherium, Xiphodon* and *Deinotherium*; **C)** *Pleistocene Edentates of Patagonia*, with *Glyptodon* and *Megatherium*.

Hawkins petitioned Owen for assistance in renewing his work at Crystal Palace and held out hope for his return even in the late 1850s, but the continued failure of the Crystal Palace Company to recoup its vast debts saw Hawkins never invited back to recommence construction on additional models. He was not even permitted to perform *pro bono* repair work in the 1870s (Bramwell and Peck, 2008). Hawkins remained attached to the project in other capacities, delivering public lectures about his prehistoric creations to the public in the late 1850s. For all its appeal and media accolade, the Geological Court did little to help the fortunes of the Crystal Palace as a whole, which continued to struggle financially throughout the following decades and never paid off its debts.

For Hawkins and Owen, the cancellation of the Geological Court project had drastically different impacts. Owen, ensconced in academic and political concerns and having barely worked on the project in the first place, probably had little regard for the fate of the models his name had been used to promote. He certainly seems to have never given them a second thought in his scholarly work, even when new discoveries supported his predicted dinosaur anatomy (*see* Chapter 7; Norman, 2000). Tellingly, while modern histories recognize the Crystal Palace Dinosaurs as a milestone in the understanding and democratization of palaeontological knowledge, the first recollection of Owen's life – an extensively detailed, two-volume biography penned by his grandson – only briefly mentions his experiences with the Crystal Palace Company and focuses mainly on the New Year's Eve banquet where Owen was the guest of honour (Owen, 1894).

Conversely, the Geological Court was far more important to the career of Hawkins: it is no exaggeration to say it was life-changing (Bramwell and Peck, 2008). Much of his post-Geological Court career revolved around the reputation he had developed as a reconstructor of extinct life. From the 1850s onward he engaged in and commanded high fees for grand public lectures. At his lecturing acme, Hawkins produced large drawings of prehistoric animals on stage (of such size that a ladder was required to reach the top of the chalkboard) to bring his restorations directly to his audience.

A strong reputation among British academics meant Hawkins enjoyed a warm reception in the US, where he was offered many auspicious platforms and opportunities by well-regarded officials and academic institutions. These included the production of (sadly never realized) Crystal Palace-like prehistoric animal installations in Central Park (see below) and the Smithsonian, as well as a collaboration with palaeontologist Joseph Leidy to create the first-ever mounted dinosaur skeleton at the Academy of Natural Sciences, Philadelphia. The skeleton, representing the then-newly discovered *Hadrosaurus foulkii*, was largely fabricated but benefitted from the discovery of complete fore- and hindlimbs, allowing Leidy to confidently infer a bipedal habit for this genus. Hawkins restored *Hadrosaurus* with an upright, somewhat kangaroo-like pose in total contrast to his Crystal Palace depictions. This different body shape must have been striking to Hawkins and made him question the many predictions about dinosaur form he had made in the early 1850s.

Hawkins continued to produce palaeoart until at least the 1870s when he completed what could be considered as his second most substantial set of palaeoartworks: seventeen paintings of different geological periods for the College of New Jersey (now Princeton University), a selection of which are shown in Figs 12.8–12.9. These artworks capture the evolution of Hawkins' personal ideas on prehistoric life and, while less seminal than his work at the Geological Court, they further track our understanding of prehistory during the late nineteenth century.

From the perspective of the Crystal Palace project, the most interesting and relevant painting of this set is the 1877 painting *Jurassic Life of Europe*, where Hawkins revisited his Crystal Palace subjects for what might have been the final time. Hawkins often referenced his Crystal Palace models in his artwork, even replicating their posing (e.g. Figs 12.4–12.6), but by 1877 he had evidently realized that many of his reconstruction choices were no longer tenable. *Jurassic Life of Europe* (a painting composed around a suspiciously familiar, Secondary Island-like landmass) shows both *Iguanodon* and *Megalosaurus* with visibly shorter forelimbs, long, bird-like hindlimbs, and relatively slender necks with somewhat smaller, more gracile heads. *Iguanodon* is also now hornless. Although still posed quadrupedally, these restorations indicate the influence of new dinosaur discoveries on Hawkins' classic dinosaur interpretations. This 1877 work might be criticized for his animals not having attained the status of true bipeds, as Hawkins realized was appropriate for the dinosaurs *Hadrosaurus* and '*Laelaps*' (= the tyrannosaurid *Dryptosaurus*) shown in other paintings of the series (*see* Fig. 12.9A). *Megalosaurus* was still considered to have powerful, bulky shoulders at this time, however (a holdover from Owen's 1856 interpretation) and, while *Iguanodon* was considered by Mantell (1848) to have relatively gracile limbs capable of non-supportive functions (e.g. grasping vegetation), an explicit case for *Iguanodon* bipedality had not yet been made. The complete Bernissart *Iguanodon* skeletons that made this conclusion unavoidable would not be found until a year after Hawkins completed this painting. There is some historic poignancy in Hawkins creating one of the last pieces of original palaeoart where overtly Mantellian and Owenian interpretations of dinosaurs retained some scientific validity.

These were to be the last palaeoartworks of note created by Hawkins before records of his career and activities taper off in the 1880s. It is thought that his last fifteen years were spent in some financial difficulty (Bramwell and Peck, 2008) and it is possible that the obsolescence of his Crystal Palace creations played some role in this. By now in his 70s, Hawkins had long ridden the wave of popularity and fame generated by his Geological Court models, but their scientific antiquation and the arrival of a new generation of palaeoartists in the late nineteenth century must have lessened his draw as an artist and speaker. His outspoken anti-Darwinian views, which often formed the topic of his later lectures (Bramwell and Peck, 2008), would only have further characterized him as a relic from another time. Hawkins became such an obscure figure in late life that he was prematurely reported as deceased by Rev. H. N. Hutchinson in his 1893 book, *Extinct Monsters*. Hawkins died in 1894 without note, and it is perhaps only now – well over a century since – that his full contribution to the popularization of palaeontology and his early mastery of palaeoart is being appreciated.

'Most successfully accomplished' or 'a gross delusion'? The response from nineteenth-century palaeontology and the impact on palaeoartistry

As Hawkins' professional presence and fame ebbed in the late nineteenth century, his models remained key attractions at Crystal Palace Park, even holding prominent positions on advertisements for the grounds in 1911. But the financial situation that halted work on the Geological Court was never rectified, meaning that its fixtures were effectively frozen in time from 1854 onwards. No updates in light of errors, new science or even substantial maintenance was permitted. This left the display vulnerable to critics who, from the moment Crystal Palace Park opened, had no qualms expressing strong opinions on the Geological Court.

Among the earliest were the learned audience of Hawkins' 1854 lecture – the same that encouraged Hawkins to reproduce his works for educational purposes – who provided a generally positive response. James Tennant praised 'the skill with which Mr Hawkins built up, piece by piece, those gigantic and extraordinary representations' (Hawkins, 1854, p. 447) and a Professor MacDonald declared the entire enterprise 'most successfully accomplished, as far as he was competent to judge' (p. 448). But amid these positive responses were comments proving that a project as ambitious as the Geological Court was never going to satisfy everyone. Evan Hopkins wondered if the Crystal Palace Company should look to demonstrating the diversity of Deep Time in other parts of the world, while Alexander Campbell thought the Geological Court had overlooked a matter of 'greater magnitude, and more lasting importance' (Hawkins, 1854, p. 449): that of our changing planet. Campbell felt that the sculptures needed contextualizing against geologically-accurate environmental conditions, as it was these that primarily dictated the sort of life that could exist on Earth. Campbell's thoughts were echoed by others, such as botanist John Lindley, who expressed concerns over juxtaposing ancient animals alongside modern plants (Secord, 2004).

Other critics were harsher. Among the most savage was the curator of the natural collections at the British Museum, John Edward Gray, a former patron of Hawkins who had worked with him on a number of projects (including the artwork shown in Fig. 2.7). Despite their familiarity, in a private letter Gray compared Hawkins' display to the showpieces of P. T. Barnum:

> The restored fossil animals are a gross delusion, Cuvier himself never attempted to give the external surface of the fossil animal, he merely gave an individual outline of what might be its general form. The models are not what they profess to be but merely enormously magnified representations of the present existing animals presumed to be the most nearly allied to the fossil – and often on very slender and sometimes on what is now known to be erroneous grounds.
> GRAY, 1854 (QUOTED IN SECORD, 2004, P. 158)

Gray's comments seem out of touch with the reality of palaeoart in the 1850s, as well as the work that had gone into the models. As demonstrated in earlier chapters, Hawkins was not merely scaling up existing species to grandiose proportions and, while Cuvier certainly avoided giving his restorations more than simple outlines (*see* Fig. 2.14), the Geological Court sculptures were hardly the first artworks to portray fully coloured and textured extinct creatures (e.g. Figs 1.8–1.9). It is difficult not to read Gray's criticism as being opposed to palaeoart in general – an attitude that was not atypical in the nineteenth century (Rudwick, 1992).

Gray's letter represents the theme of subsequent discussion of the Geological Court, where the geological components of the displays were largely forgotten by popular and academic authors and conversations focused entirely on the restored animals. Scientific accuracy was only one

Fig. 12.10 Accusations that Hawkins' sculptures were 'enormously magnified representations of the present existing animals' are unfounded when compared against his sculptures. Reconstructions like his *Mosasaurus* are obviously heavily influenced by living species, but are unique creations guided by fossil data, not merely scaled-up modern animals. (2021)

of the concerns raised about the display: what were visitors actually learning from this array of fabricated geological features and restored extinct animals that lacked even the most basic interpretive signage? The sociologist Harriet Martineau gave some perspective on this in an 1854 article for *The Westminster Review*, noting that the Geological Court was enormously popular, but that most visitors had little idea what the animals represented. Most, at best, left with the idea that they were looking at 'antediluvians' – animal species that went extinct in the Biblical Flood (Martineau, 1854). This might, perhaps, be seen as an educational victory given the relative novelty of the concept of extinction among the public in the early 1850s, but it fell well short of communicating the complex geological and palaeontological information infused into the Geological Court by Ansted, Campbell, Hawkins and Owen. As time passed, the outright educational value of the displays were seen to diminish and they even became the subject of mockery, especially as they fell into disrepair (Secord, 2004). Writing in 1894, Owen's grandson questioned their value to the general public and highlighted their obvious age and dilapidation:

> The writer lately made a pilgrimage to the Crystal Palace and succeeded in effecting a surreptitious landing upon the island where the forms of these extinct monsters are displayed. Here he found the specimens in question slightly dilapidated as to tails and other extremities, together with a total absence of anything like explanation, or even names of the creatures. From the remarks of the British holiday-makers he gathered that the popular mind was divided as to whether these images were inferior imitations, on a large scale, of certain animals at the Zoological Gardens – wherein the popular mind had a vague sense of being defrauded – or whether they were not creations of some eccentric person's imagination. One individual was of opinion [sic] that they were surely placed there with the pious purpose of setting clearly before the eyes of the public, as a terrible warning, the fantastic visions sometimes seen by such as are in the habit of indulging too freely in spirituous liquors.
>
> OWEN, 1894, PP. 398–99

The initially positive academic response achieved by the Geological Court had now turned consistently fierce, especially concerning the three dinosaur species. It is not hard to see why this was the case. *Iguanodon*, *Megalosaurus* and *Hylaeosaurus* were the star attractions of the Geological Court and their heavyset, quadrupedal forms loomed large in the press and collective consciousness. Science had already established the basic shapes of creatures like ichthyosaurs and plesiosaurs before the 1850s, but our knowledge of basic dinosaur anatomy was still developing: the most celebrated components of the Crystal Palace display were highly vulnerable to new data and improved interpretations.

Comparisons with revised reconstructions became inevitable as newly unearthed fossils found in 1860s and 1870s Europe and North America left little doubt about the reality of dinosaur body shapes. This material, including the armoured dinosaur *Scelidosaurus*, the predatory '*Laelaps*', the herbivorous *Hadrosaurus*, and complete specimens of *Iguanodon*, had direct bearing on the animals restored at Crystal Palace. As noted in Chapter 7, the canny anatomical insights of Mantell, Owen and Hawkins meant that these new discoveries did not reinvent *everything* we knew about dinosaur life appearance, but they certainly exposed the gulf between Hawkins' predicted forms and real dinosaur anatomy.

The embarrassment of Hawkins' dinosaurs was not missed by palaeontologists and educators. The fact that many of his other restorations were holding up well against changing science was overlooked and the Crystal Palace Dinosaurs were used as examples of erroneous, misleading reconstructions and poor scientific outreach. The January 1884 edition of *Popular Science Monthly* reproduced an illustration of a Hawkinsian *Iguanodon* specifically to demonstrate what this animal *didn't* look like, describing it as 'an illustration of the danger of making too hasty generalizations from too few or too imperfectly understood data' (p. 353). In *Geschichte und Methode der Rekonstruktion vorzeitlicher Wirbeltiere* ('History and method of reconstruction of ancient vertebrates'), one of the first major reviews of palaeoartistry, the Austrian palaeontologist Othnieo Abel launched an especially harsh attack on Hawkins for having insufficient scientific training and compared his *Iguanodon* to pre-1800 artwork where fossil vertebrates were interpreted as unicorns, giants and dragons (Abel, 1925).

Just as scathing were comments by the revered American palaeontologist Othniel Marsh. Marsh was among the researchers who had, by virtue of a bounty of dinosaur and other fossil reptile discoveries in the American midwest, advanced our understanding of dinosaurs considerably from the mid-nineteenth century views expressed at Crystal Palace. He was responsible for discovering and naming

Fig. 12.11 Hawkins' concept sketch of the never realized New York Paleozoic Museum, 1868. This indoor enterprise would have shown different and seemingly more dynamic displays than the Geological Court, including the tyrannosauroid '*Laelaps*' (= *Dryptosaurus*) harassing *Hadrosaurus* next to a plesiosaur (possibly *Elasmosaurus*), two *Megatherium*, two *Glyptodon*, and extinct cats and elephants. Many of these ideas were recycled in Hawkins' 1870s oil paintings for the College of New Jersey (Figs 12.8–12.9).

iconic taxa like *Allosaurus*, *Ceratosaurus*, *Brontosaurus* and *Diplodocus*, but he had little interest in life reconstructions, believing that our understanding of extinct animals was insufficient for this purpose (Dodson, 1996). This conservative attitude was underpinned by concern for public education, as he believed erroneous reconstructions were difficult to remove from public consciousness once established. An 1875 private letter concerning Hawkins' '*Labyrinthodon*' outlined his general thoughts on the field of palaeoart:

> I do not believe it possible at present to make restorations of any of the more important extinct animals of this country that will be of real value to science, or the public. In the few cases where materials exist for a restoration of the skeleton alone, these materials have not yet been worked out with sufficient care to make such a restoration perfectly satisfactory, and to go beyond this would in my judgment almost certainly end in serious mistakes. Where the skeleton, etc., is only partly known, the danger of error is of course much greater, and I would think it is very unwise to attempt restoration, as error in a case of this kind is very difficult to eradicate from the public mind… A few years hence we shall certainly have the material for some good restorations of our wonderful extinct animals, but the time is not yet.
>
> MARSH, 1875 (QUOTED IN DODSON, 1996, P.74)

This broader philosophizing provides context for Marsh's more direct addressing of the Crystal Palace restorations in 1895, when visiting the British Science Association:

> The dinosaurs seem… to have suffered much from both their enemies and their friends. Many of them were destroyed and dismembered long ago by their natural enemies, but, more recently, their friends have done them a further injustice by putting together their scattered remains, and restoring them to supposed lifelike forms… So far as I can judge, there is nothing like unto them in the heavens, or on the earth, or in the waters under the earth. We now know from good evidence that both *Megalosaurus* and *Iguanodon* were bipedal, and to represent them as creeping, except in their extreme youth, would be almost as incongruous as to do this by the genus *Homo*.
>
> MARSH, 1895 (QUOTED IN DESMOND, 1976, P. 41)

These critical reactions to the Geological Court have been framed as one reason why few efforts were made to recreate ancient life on this scale again throughout the late nineteenth century, despite the zeal for grand exhibitions of international science and industry at this time (Secord, 2004). There may be some truth to this, although Nieuwland (2019) highlights the importance of another factor in this lack of copycat exhibitions: nationalism. Interest in extinct life tended to follow local discoveries (Nieuwland,

Fig. 12.12 Scenes from the *Anthropological Exhibition of Moscow*, 1879, featuring prehistoric animals inspired by those of the Geological Court. The plesiosaur, mammoth and Kuwassegian *Iguanodon* in these photos were reportedly capable of moving via a mechanism of wires and gears.

2019) and most countries were lagging behind Britain in identifying a range of fantastic fossil animals. Merely replicating the content of the Geological Court abroad would neither satisfy cultural and political goals nor, as was rapidly becoming apparent, scientific ones, and no-one wanted an embarrassment on their hands like the already obsolete British reconstructions.

The temptation to create something like the Geological Court was palpable in the United States, however. P. T. Barnum expressed interest in having Hawkins create such a display (Secord, 2004) and, in 1867, work began on the never-realized Palaeozoic Museum of Central Park, New York. This was Hawkins' second go at an exhibit on the scale of the Geological Court and featured updated dinosaur reconstructions based on American finds, including bipedal forms of *Hadrosaurus* and '*Laelaps*', alongside other prehistoric animals in more dynamic poses than achieved at Crystal Palace. Corrupt politics and spiralling expenses saw the project cancelled with a vengeance, however, culminating with the infamous Tweed gang hiring thugs to smash the half-completed models with sledgehammers (Desmond, 1974; Bramwell and Peck, 2008). Only relatively limited large-scale efforts to reconstruct extinct animals followed in the United States until the twentieth century, where growing interest in dinosaurs – driven by spectacular discoveries and another leap ahead in palaeoart proficiency and visibility – saw grand, outdoor recreations of extinct life realized for the first time on US soil.

On the other side of the world, Russian zoologist Anatoli Bogdanov and his colleague, D.N. Anuchin, achieved a rare success at replicating some aspects of the Geological Court as part of the *Anthropological Exhibition of Moscow*, 1879. Although primarily focused on Russian anthropology, this indoor exhibit was directly inspired by visits to the Crystal Palace (Knight, 2001) and featured life-sized models of several prehistoric animals. Photographs of the exhibition show a plesiosaur, ichthyosaur, dinosaur (probably *Iguanodon*), mammoth and glyptodont within a stereotyped 'prehistoric' landscape, and other animals – likely pterosaurs and some sort of Devonian 'amphibian' – are also mentioned in a November 1879 *Theosophist* account. The same write-up also mentions that 'clever mechanisms of wires, wheels and springs' allowed the models to move – surely among the very first examples of animated prehistoric creatures of any format. Their technical execution was not matched by their anatomical insight, however, as the models were based on

Fig. 12.13 A selection of artworks influenced by the Geological Court and related Hawkins palaeoart. **A)** Charles H. Bennett's cartoon *Megalosaurus* as seen in John Cargill Brough's 1858 *The Fairy Tales of Science: A Book for Youth*; **B)** Secondary and Tertiary animals from Ansted's 1863 *The Great Stone Book of Nature*; **C)** Edward Riou's Wealden Scene from Louis Figuier's 1863 book *La Terre avant le déluge*, featuring a Kuwassegian *Iguanodon* and Owen's 1854 guidebook *Megalosaurus*; **D)** one of several plates based on Hawkins' posters from Margaret Plues' 1863 *Geology for the Million*.

restorations that were by then decades old. The *Iguanodon* in particular is striking for evidently being based on Josef Kuwasseg's c. 1850 artwork, complete with small, crouching limbs and lizard-like proportions (*see* Fig. 1.9F). Though popular within Russia and seemingly innovating the technological portrayal of extinct life, the Moscow Anthropological Exhibition had little, if any, impression on international palaeontology or palaeoart.

While interest in grand displays of prehistoric animal sculptures struggled to find footing and impact in the late nineteenth century, the Crystal Palace Dinosaurs left a more noticeable impression on two-dimensional palaeoart. Drawings and paintings of extinct life continued production with the same zeal as seen pre-1854 and many now sported distinctly Hawkinsian flourishes (*see* Fig. 12.13). The practice of copying existing reconstructions was commonplace in the 1800s such that, for all the abundance of nineteenth-century palaeoart, only a fraction contained genuinely original reconstructions. It was the fate of any new and influential palaeoartwork to be copied again and again by other

Fig. 12.14 The evolution of dinosaurs in two-dimensional palaeoart, post-Crystal Palace. **A)** Edward Drinker Cope's 1869 scene of American Cretaceous reptiles: *'Laelaps'* confronts *Elasmosaurus* with *Tylosaurus* behind; the turtle *Osteopygis* swims in front of *Hadrosaurus* and *Thoracosaurus* on the shore; **B)** scene from J. W. Dawson's 1874 *The Story of the Earth and Man* of 'herbivorous dinosaur' and 'labyrinthodont', with carnivorous dinosaurs in the background; **C–D)** images from Flammarion's (1886) *Le Monde Avant la Création d'Homme*, with bipedal *Megalosaurus* and *Iguanodon* (C) and a New World *Stegosaurus* (left) meeting an Old World, Kuwassegian *Iguanodon* (right) (D); **E)** a mix of Copeian and Hawkinsian dinosaurs from Fr Rolle's 1886 *G.H. von Schubert's Naturgeschichte: Geologie und Paläontologie*.

artists, and the famous Crystal Palace Dinosaurs, despite their quickly identified scientific faults, were no exception. The palaeoartistic impact of the Crystal Palace Dinosaurs can be traced most readily through copies of the distinctive dinosaur and 'Labyrinthodon' restorations but, counterintuitively, it was not the sculptures that were copied most widely. Rather, it was the *Megalosaurus* illustration from Owen's guidebook and the content of Hawkins' 1860s poster series that became most prevalent, presumably because these could be replicated without a dedicated visit to London to sketch the models directly. Derivative artworks appeared in too many places to list, but examples shown in Figs 12.13 and 12.14 include works from John Cargill Brough's 1859 *The Fairy Tales of Science: a Book for Youth*, Fr. Rolle's 1886 *G.H. von Schubert's Naturgeschichte: Geologie und Paläontologie*, Margaret Plues' 1863 *Geology for the Million*, Ansted's 1863 *The Great Stone Book of Nature*, and Louis Figuier's 1863 *La Terre avant le déluge*.

The adoption of Crystal Palace-based restorations into so many artworks firmly embedded them into the nineteenth-century 'canon' of prehistoric depictions, although they did not obscure the influence of older pieces: Goldfuss' 1831 *Jura Formation* (see Fig. 1.8B) and Josef Kuwasseg's *c.* 1850 take on *Iguanodon* (see Fig. 1.9F) remained widely referenced. By the second half of the 1800s artists were often combining different reptiles from these sources into the same compositions (e.g. Figs 12.13–12.14), inadvertently mixing palaeontological interpretations spanning several decades. This trend continued despite new information and skeletal reconstructions based on superior fossil materials trickling into books and scientific papers (see Fig. 12.15). Amazingly, Hawkinsian dinosaurs were still featured in

Megalosaurus **skeletal and life reconstruction, 1893**

Iguanodon bernissartensis **skeletal and life reconstruction, 1893**

Fig. 12.15 It was not until the end of the nineteenth century that European palaeoartists began revising the life appearance of dinosaurs in earnest. The 1893 work of Dutch artist Joseph Smit for Henry Neville Hutchinson's book *Extinct Monsters: A Popular Account of Some of the Larger Forms of Ancient Animal Life* was among the earliest efforts to reconstruct these species post-Crystal Palace, replacing Hawkins' dinosaurs with almost unrecognizable new body plans.

artwork as scientifically legitimate reconstructions on these animals as late as the 1880s (see Fig. 12.14C–E) and, in at least one instance, the early 1900s (Nieuwland, 2019). Some artworks, presumably produced by naive artists, justified copying the vintage Hawkinsian dinosaurs with new American-produced art of bipedal forms as reflecting a geographic distinction: European dinosaurs were more 'reptilian' and quadrupedal than their kangaroo-like American relatives (e.g. Flammarion, 1886). Whatever the justification, it's hard not to see this as vindication of Marsh's concerns that erroneous depictions of prehistoric life would prove hard to remove from public consciousness – an issue still troubling educators and science communicators today.

A second, almost paradoxical impact of the Crystal Palace Dinosaurs on two-dimensional palaeoart was to seemingly curb enthusiasm for novel palaeontological reconstructions for several decades (Secord, 2004; Nieuwland, 2019). A number of new, highly-restorable reptile species (e.g. *Scelidosaurus, Hypsilophodon, Archaeopteryx, Pliosaurus*) and superior examples of known genera were discovered and studied in the decades following the opening of the Geological Court but restorations of them were few and far between. Skeletal restorations were commonplace, as were recycled palaeoartworks from the 1850s or before, but 'corrected' life reconstructions of animals like *Iguanodon* would not appear widely until the 1880s (see Figs 12.14–12.15; Flammarion, 1886; Hutchinson, 1893). Even new and incredibly exciting animals like the sauropod dinosaurs – known from good and exceptional skeletons uncovered in the UK and US from 1871 onwards (Taylor, 2010) – did not receive flesh restorations until the end of the century (Flammarion, 1886).

Given the readiness to depict newly discovered extinct animals in the decades preceding the Crystal Palace display, we must wonder if the opinions expressed by Marsh, Gray and similarly minded individuals about the Geological Court stifled the development of novel palaeoartworks for a generation. It must, admittedly, have been difficult for late-nineteenth-century palaeontologists to see rapidly outdating mid-Victorian palaeoart without viewing it as ill-judged, over-confident media that could hamper public education more than helping it.

This is not to say, however, that absolutely no novel, non-derivative palaeoartworks were produced following Crystal Palace. From the late 1860s onwards American palaeoartistry began to take shape, with early works showing new interpretations of marine reptiles (plesiosaurs and mosasaurs), pterosaurs and dinosaurs as revealed by new, often excellent fossils from the central and western United States (see Fig. 12.14A; Dawson, 1973). As noted above, Hawkins contributed to the rise of American palaeoart, producing oil paintings of extinct animals for the College of New Jersey to create one of the few reasonably sizeable palaeoart projects of this era. American palaeoart would, from the 1890s onward, lead the development of the genre for the next century, and we have to wonder if the reaction to the Crystal Palace Dinosaurs and the seeming timidity of European artists to experiment with new depictions of extinct species played into this. The irony of the Geological Court being a reference source for naive artists of prehistoric animals for several decades, while also potentially dampening development of new European palaeoart for the same duration, is difficult to avoid.

Reappraisal and modern recognition

From the 1890s onward scholarly views on the Geological Court began to soften. Several decades' worth of hindsight allowed a new generation to see that the Geological Court had to be evaluated with appropriate historic context and not, as perhaps some of its earlier critics had done, measured against the science of subsequent decades. It's perhaps these more nuanced perspectives, and not all-out criticism, that have become the most enduring scholarly opinion of the sculptures.

One of the earliest defenders of the Crystal Palace Dinosaurs was Henry Neville Hutchinson who, in his 1893 book *Extinct Monsters: A Popular Account of Some of the Larger Forms of Ancient Animal Life*, praised Hawkins' work even when highlighting their deficiencies against later science. *Extinct Monsters* was full of new palaeoart by the Dutch artist Joseph Smit (see Fig. 12.15) and he may, therefore, have been more sympathetic to both the difficulty of reconstructing fossil animals while also seeing their educational value. He discussed all of Hawkins' dinosaurs with their 'very excusable mistakes' and repeatedly mentions the challenge of restoring ancient animals from incomplete remains, encouraging readers to judge the Crystal Palace through an appropriate historic lens:

> Visitors to Sydenham, who have wandered about the spacious gardens so skillfully laid out by the late Sir Joseph Paxton, will be familiar with the great models of extinct animals on the 'geological island'. These were designed and executed by the clever artist, Mr. Waterhouse Hawkins, who made praiseworthy efforts

to picture to our eyes some of the world's lost creations, as restored by the genius of Sir Richard Owen and other famous naturalists. His drawings of extinct animals may yet be seen hanging on the walls of some of our provincial museums' but many of these are far from correct. The difficulties were much greater in those early days.

HUTCHINSON, 1893, P. 34

Hutchinson's 1894 follow-up, *Creatures of Other Days*, focused less on subjects covered at Crystal Palace but still offered a pragmatic view of Hawkins and his dinosaurs, and drew attention to the greater accuracy of the non-dinosaurian models:

These [models], although some of them were made from very imperfect materials, such as did not really warrant any attempt at restoration, nevertheless are interesting, historically, as being the first attempts of the kind, and at least serve to draw the attention of those who visit the Palace to the wonders of geology... the huge *Iguanodon*... and the ponderous *Megalosaurus* next to it are very far from the truth, but some of the others further on are much more in accordance with modern ideas on the subject.

HUTCHINSON, 1894, P. 143

Elsewhere in Europe, Hans Becker's 1911 article *Alte und neue Rekonstruktionen ausgestorbener Tiere* ('Old and new reconstructions of extinct animals') presented a discourse on the changeable nature of prehistoric animal reconstructions that echoed Hutchinson's sentiments. Becker noted with scientific objectivity that restorations are prone to change when new information becomes available and used Hawkins' *Iguanodon* to demonstrate the transient nature of palaeoartistry rather than as an example of an outdated, erroneous reconstruction to be ridiculed. Becker even stressed the role that imagination and prediction play in the formulation of palaeoart, and ended his article with the suggestion that new finds will eventually warrant revision to the contemporary artworks it contained. Views such as this suggest a level of maturity and perspective about palaeoartistry lost on earlier critics of the Crystal Palace project.

By this point, the Crystal Palace Dinosaurs were well over half-a-century old and their role in palaeontological outreach had diminished. New authors and palaeoartists – most obviously Charles Knight (Milner, 2009) – as well as new natural history museums were helping to modernize the public's interpretation of prehistory. Visitors to Crystal Palace Park, which was now approaching its final years of opening, would have seen the Geological Court as a series of visibly ageing displays from another scientific era, allowing Hawkins' sculptures to be recast in a historic perspective.

Fig. 12.16 The lower portion of the Secondary Island, 2015. This view captures a range of scientific credibility in Hawkins' models: almost entirely speculated animals (*Dicynodon* and '*Labyrinthodon*'), part-reconstructed, part-speculated animals (dinosaurs) and well-informed, relatively accurate animals (enaliosaurs and *Teleosaurus*). Much of the nineteenth-century response to Hawkins' work overlooked this nuance, however, focusing more on his failures than his successes.

Rather than representing outdated restorations threatening public education, they were now artefacts from a previous generation of scientific interpretation. But, more often, their growing obsolescence saw them simply forgotten. Their relevance to discussions of prehistoric life long faded, they were rarely mentioned in popular palaeontological texts and, when brought up, were mere footnotes.

The fire that destroyed the Crystal Palace in 1936 and saw the repurposing of its grounds as a public park (see Chapter 13) further redefined the role of the Geological Court, mostly by pushing it further into obscurity. Now in the ownership of local councils and no longer a destination for holidaymakers and pleasure seekers, the ageing geological and palaeontological displays became a curiosity known only to locals and those of specialist historic interests. The enduring charisma of Hawkins' sculptures has ensured they remain a valued part of the London landscape, however (Clayton, 1968), and their historic importance has been recognized by major authorities since 1973 when they were classed as Grade II Listed Buildings on Historic England's National Register of Heritage Monuments. This was upgraded to Grade I in 2007 and elevated again in 2020 when they were placed on Historic England's register of 'At Risk' structures, marking them as highest priority for conservation.

Modern scholars and palaeoartists have also not forgotten the significance of the Crystal Palace Dinosaurs. Although some retellings of their history have been over-simplistic or more Owen-centric than recent research suggests is appropriate, they have still become a celebrated milestone in accounts of palaeontological history. The many works mentioning them include passages standing up to their nineteenth-century critics, highlighting their lack of historic consideration and questioning whether anyone could have improved on Hawkins' achievements (Desmond, 1976; Norman, 1991).

> [The Crystal Palace Dinosaurs] were a brilliant innovation in their time, based as they were on extremely poor fossil material, and it is unlikely that anyone, especially Marsh, could have come up with anything better at the time.
>
> NORMAN, 1991, P. 57

Gregory S. Paul, widely regarded as one of the most influential palaeoartists of modern times, described Hawkins as 'the first significant artist to apply his skills to the field' of dinosaur art (Paul, 2000). In their overviews of pre-modern palaeoart, both Martin Rudwick (1992) and Zoe Lescaze (2017) recognize Hawkins and his Crystal Palace work as the democratization of palaeontology and an important step in encouraging scientific interests among laypeople. Hawkins' biographers, Valerie Bramwell and Robert M. Peck (2008), have made similar points and credit Hawkins with preparing the public for the debates on Darwinian evolution that shortly followed the opening of the Crystal Palace. Dinosaur television documentaries have also shown continued interest in the Geological Court. A comment in Tom Holland's 2011 BBC television documentary *Dinosaurs, Myths and Monsters* about his gradual acceptance of Hawkins' displays, despite their inaccuracy, encapsulates 170 years of discussion and criticism into a few sentences:

> ...when I was a child, I made a point of refusing every offer from my parents to take me to Crystal Palace. These reconstructions offended every last bone in my dino-geek body. But now that I'm here, I can realize what a little prig I was being. This [*Megalosaurus*] built of concrete may not be cutting-edge palaeontology, but it tells you everything about why dinosaurs still fascinate us.
>
> TOM HOLLAND, 2011,
> *DINOSAURS, MYTHS AND MONSTERS*

Reflecting on a complex legacy

It is difficult to wrangle all these views and historic points together in a manner that adequately summarizes the impact of the Crystal Palace Dinosaurs, especially because virtually all opinions expressed on them – even the most critical – have some legitimacy. Marsh's rejection of palaeoart and Hawkins' work, for example, was certainly po-faced and strongly worded but his underlying concerns – that the fossil material for some of Hawkins' species was inadequate to inform a reconstruction, and that erroneous restorations would be difficult to remove from the public eye – were entirely validated. Conversely, Hutchinson and others were also right in drawing attention to the value of the Geological Court as an inspirational and educational place that encouraged interest in geological topics and also, among the less accurate models, contained many fine, scientifically credible palaeoartworks.

There is a temptation when reviewing historic topics to distill reflection down to the simple question of whether a project was a 'success', but the Crystal Palace Dinosaurs defy such easy categorization. It may be enough to simply reiterate that their impact, both positive and negative, was significant, and that they represented the first time the applications of palaeoart – as a crowd-drawing spectacle, as an educational tool, as a commercial enterprise – were fully realized in a single project.

To appreciate this point further, it is worth exploring how the palaeoartistic practices and principles of scientific communication pioneered in 1854 are still largely played out today, and that the first discussions catalysed by the Crystal Palace Dinosaurs continue even now. We still, for instance, reconstruct life-size models of poorly known fossil animals to high levels of detail despite the amount of speculation and inference this requires, a practice that is still questioned by some (Hallett, 1987; Witton, 2018). Our lack of caution around reconstructing poorly known fossil taxa includes several major palaeoartistic celebrities: many of the most popular giant pterosaurs and dinosaurs are actually very poorly represented by fossils, and yet they dominate the modern palaeoart canon.

Similarly, palaeontological studies based on provisional or controversial data are still the basis of major palaeoartworks and publicity campaigns that shape popular opinion, only to be strongly criticized, and in some cases entirely overturned, by fellow researchers largely out of the public's view, in academic discourse. We still have the same lack of reservation for the mass publicizing of new, tentative data that Hawkins had when creating Owen's reptilian frogs. And Marsh's prediction that inaccurate or outdated reconstructions have lasting impacts on public education has been demonstrated again and again, from the seemingly indefatigable presence of now-outdated 1993 *Jurassic Park*-style dinosaurs in palaeontological media to studies showing that twenty-first-century children still mostly imagine dinosaurs with upright, kangaroo-like postures (Ross *et al.*, 2013). In this light, any judgement we have of the Crystal Palace Dinosaurs as overzealous palaeoartistry and science communication is hypocritical: they serve less as a cautionary tale and more as the first realization of what would become typical practice in palaeontological outreach, the merits and detriments of which are as much a matter of discussion today as they were almost two centuries ago.

It thus seems most prudent to simply appreciate the Crystal Palace Dinosaurs as what they are: ambitious, well-executed and pioneering works of palaeoartistry, for all the good and bad that such artworks represent. To criticize them is to criticize palaeoartistry in general. The fact that they capture a snapshot of time when extinct life was only partly understood is now a major element of their charm and appeal, and grants them a unique, highly significant place in history. They are not a physical commemoration of failure but an inspiring monument to the ever-changing and self-correcting scientific process, as well as a testament to how pioneering scientists and artists were able to understand much about Deep Time and ancient life with only rudimentary investigative techniques and inferior fossil data. If we focus on what Hawkins, Ansted, Campbell and Owen got right, rather than what they got wrong, and contextualize what they produced against the information they had to work with, their achievements were nothing short of remarkable.

Finally, we can also take a view which ignores any concerns about the scientific credibility or public impact of the Geological Court, and simply appreciate it as a collection of amazing artworks of animals and geological features in its own right. The survival of the Geological Court displays in Crystal Palace Park been a privilege for the many that have visited them and we should not take their continued existence for granted. Had they not endured for the last 170 years, they would surely be the stuff of palaeontological legend: a lost display from an early, explorative era of earth sciences that future generations would dream of experiencing first hand.

CHAPTER 13

Past becomes future: the conservation of the Geological Court

Visitors to the Geological Court versed in the history of the site may consider themselves lucky that the bulk of the creative genius, ambition and educational spectacle developed by Ansted, Campbell, Hawkins and Owen has survived more or less intact to the modern day. And yet, luck has only played a minor role in the continued residence of prehistory in south London: it is generations of park custodians, conservators, scholars and volunteers who have kept the sculptures, Geological Illustrations and landscaping repaired and in sufficient condition to ensure their longer term survival. The history of maintenance at the site, it must be said, is not spotless. Although paint analysis suggests repainting may have been a regular occurrence – perhaps at a typical rate of every six years (Hirst Conservation, 2000) – many components of the Geological Court have been badly neglected, vandalized and even intentionally destroyed in the last 170 years. It is largely through a series of punctuated conservation projects, rather than continued maintenance, that the essence of the original Court can still be enjoyed today.

A number of issues contribute to the degradation of the Crystal Palace Dinosaurs that must be addressed more or less continuously to keep the site in good order (Brierley et al., 2018). Virtually all of these relate to two factors: 1) its status as an outdoor display in a park setting with a turbulent history; and 2) the unconventional assemblage of materials used to build structures of unusual design and purpose. Combating the many factors degrading the Geological Court is neither easy nor cheap, and without Hawkins' and Ansted's notebooks, or even their business records, we don't know much for certain about the construction of the sculptures: conservators only learn about these aspects through investigation and experimentation. Keeping the Geological Court in order is thus a conundrum handed down from generation to generation, requiring as much research and learning about the site as physical repair work.

The exposure of the models to all weathers and seasons subjects them to a familiar set of damaging agents: fluctuating temperatures, precipitation, and occasional extreme conditions. The latter, in the form of storms, strong winds and heavy snowfall, provide short-term but harmful stresses to delicate parts of the palaeontological sculptures. Other effects are less dramatic but no less severe. Daily changes in temperature see the different materials within the displays expand and contract at varied rates, disrupting and weakening their internal structure.

Water – both in the Tidal Lake and as rainfall – corrodes metal components and introduces problems such as freeze-thaw, where water soaked into concrete freezes and expands to crack the surrounding cement, in turn allowing more water and ice deeper in the structure. This is a perpetual

Fig. 13.1 Storms and extreme cold are among the many meteorological conservation risks to the Geological Court, the latter leading to freeze-thaw weathering and ingress for other destructive agents. **A)** *Megaloceros* in the snow (2010) and **B)** plesiosaurs, *Teleosaurus* and broken Oolite pterosaurs, knocked over by wind or vandals (2005).

Fig. 13.2 The unfortunate reality of the Crystal Palace Dinosaurs is that the Geological Illustrations and palaeontological sculptures are often overgrown and in a state of disrepair. **A)** Coal Measures covered with plants whose roots dislodge the facade of historic outcrops and destabilize the underlying foundations; **B)** Chalk pterosaurs with missing jaws and corroding wings; **C)** the broken jaw of *Ichthyosaurus communis*; **D)** *Megaloceros* stag with snapped antlers, weathered paint and missing tail; **E)** headless *Plagiolophus*; **F–G)** large cracks in the body of *Iguanodon*, indicative of structural failure; **H)** totally overgrown and fracturing *Temnodontosaurus*. All images from 2021 except B from 2018 and F–G from 2016.

risk for the partially submerged enaliosaurs, *Mosasaurus* and *Teleosaurus*, where freeze-thaw has a long history of destroying their limbs and bodies. Damp conditions also enhance the corrosive effects of pollutants, especially sulphur dioxide and chloride, on ferrous metal elements. Corrosion can split and fracture overlying concrete and, with hoop iron embedded in many sculptures, large-scale structural failure can result.

The environment of the Geological Court itself also brings risks to the displays. Seemingly once a fairly open setting, large trees, shrubs, wildflowers, mosses and lichens are now ubiquitous in Crystal Palace Park and, while clearly enjoyable to look at, environmentally beneficial and adding a suitably untamed, 'primordial' atmosphere to the Geological Court, this floral community also obscures and damages the displays. Unchecked plant growth can entirely hide sculptures and Geological Illustrations (*see* Fig. 13.2A, H), root into and undermine foundations, and colonize cracks and fissures, splitting them apart as they grow. Conservation work regularly reveals a network of roots propagating through the sculptures and Geological Illustrations and, over time, the lack of simple, routine plant maintenance can incur significant damage.

A human element of destruction has also plagued the Geological Court since the Crystal Palace Park first opened. Vandalism has long been a problem, both accidental and intentional, where displays are climbed on, graffitied, smashed and stolen. Tremendous harm can result, ranging

- **2021:** Installation of swing bridge to Secondary Island for reliable, secured access for conservation and research.
- **2020:** Geological Court added to "Heritage At Risk" Register.
- **2016-17:** Conservation of select reptile sculptures and Secondary Island by Cliveden Conservation and Skillington Workshop. Renovation of northeast weir sees inadvertent removal of several tonnes of Greensand Geological Illustration.
- **2013:** *Friends of Crystal Palace Dinosaurs* formed to promote conservation and interpretation of Geological Court.
- **2007:** Geological Court Grade I listed by English Heritage.
- **2005:** Oolite pterosaur replicas destroyed and removed from site.
- **2001-2:** Major conservation works across entire site by Morton Partnership. One *A. commune* and both Oolite pterosaurs replaced with fibreglass replicas. Reconstruction of lost Geological Illustrations. Surviving mammal sculptures returned to Tertiary Island.
- **1992:** Crystal Palace Park Zoo no longer active on Tertiary Island.
- **1986:** Park management and ownership shifts from Greater London Council to Bromley Council.
- **1976:** *Hylaeosaurus* head is replaced with fibreglass replica, original begins status as separate display elsewhere in the park.
- **1973:** Geological Court Grade II listed by English Heritage.
- **1965:** Park ownership and management shifts to Greater London Council.
- **1960-64:** Construction of Crystal Palace National Sport Centre immediately north of Geological Court. Major landworks included redesigning of rockery and waterways in front of Primary Geological Illustrations. Over 50% of the Primary strata are destroyed, including parts of cave and mine.
- **1960:** *Pets Corner* expanded into larger zoo.
- **1959:** Conservation of palaeontological sculptures supervised by William Elgin Swinton from British Museum Natural History.
- **1958:** Most recent evidence of *Palaeotherium magnum* sculpture and original *"P. minus"* head.
- **1953:** The petting zoo *Pets Corner* is opened on the Tertiary Island, displacing mammal sculptures and Geological Illustrations.
- **1952:** Park reopened to the public, ownership transferred to London County Council.
- **1936:** Crystal Palace burns and the grounds are closed to the public. Control of the park transfers to the Ministry of Defence and the sculptures are rumoured to be used for target practice. Original *"Plesiosaurus" macrocephalus* head separated from body around this time.
- **1913:** The Crystal Palace and grounds become property of the nation.
- **1911:** Ownership transfers to Robert Windsor-Clive, 1st Earl of Plymouth as a precursor to public ownership.
- **1909:** Crystal Palace Company declares bankruptcy for second time.
- **1894:** Owen's grandson reports sculptures are "slightly dilapidated as to tails and other extremities"
- **1887:** Crystal Palace Company declares bankruptcy for first time.
- **1875:** Hawkins' unhappy with dilapidated condition of site and repositioning of models around Geological Court. Volunteers to do repairs, but offer is not accepted.
- **1874:** Probable last year of artificial 'tide' functioning in Tidal Lake.
- **1866:** Fire destroys natural history and ethnological exhibits in the Palace nave, which are not replaced.
- **1855:** Last sculptures and Geological Illustrations competed, Crystal Palace Company cancels further development of Geological Court.
- **1854:** Crystal Palace Park opens in June.
- **1852-54:** Construction of the 20 acres of Geological Court. Building of sculptures and Geological Illustrations.

Fig. 13.3 A concise history of known events and conservation works related to the welfare of the Geological Court. Several obvious gaps exist in our knowledge of such activities. The decline of the Crystal Palace project makes way for a turbulent mid-twentieth century of park ownership and development before the important heritage of the Court is recognized in the 1970s. Although conservation activities became more focused thereafter, maintenance remains erratic, leading to pulses of large-scale conservation interventions.

from toothless animals and leafless cycads to broken and missing sculptures. Such damage must be repaired quickly, always at great expense, to prevent development of secondary issues related to the exposure of metal armatures or further loss of original sculpture elements.

Historic accounts date the first of these destructive acts to within months of the park opening, when visitors invaded the Secondary Island to steal the sculptures' metal teeth (Martineau, 1854). Shortly after, in 1855, *The Observer* reported visitors climbing and jumping on the palaeontological sculptures, fracturing one *Teleosaurus* jaw in the process. So significant was this threat that a note was eventually inserted into Owen's guidebook asking visitors not to damage the displays (Secord, 2004) but, even so, their extremities and tails were in poor condition by the 1890s (Owen, 1894). Then, as now, fences and the Tidal Lake are deterrents to accessing the displays, but they are easily circumvented by determined trespassers, especially in warm summer months when the lake waters – now much shallower thanks to sediment accumulation and plant growth – fall to easily traversable depths.

Another major cause of harm, custodial neglect and mistreatment, also dates back to the early years of the Crystal Palace Park (*see* Fig. 13.3). These threats concern decisions made about the Geological Court that have been insensitive to its scientific and cultural significance, allowing components to be destructively modified to make way for new park developments or repaired with little regard for their original designs and intent. Among the earliest accounts of such issues were reported by Hawkins when, after returning to England in 1874 from a visit to the United States, he found the carefully planned Geological Court in disarray after rearrangements by the Crystal Palace Company. He remarked in a letter:

> …immediately after my return to England from the U.S. America (26th July, 1874), I hastened to the Crystal Palace, to see the condition of my Colossal Models… in a sadly depreciated condition… In the limestone in the upper portion of the section there was originally an elaborate model of a Lead Mine, now obliterated with white wash; and at the present time the beds and bands of coal are disintegrated and destroyed by the ingrowth of vegetables, that has been allowed to attain the size of bushes; and to complete the inconsistency and solecism, the gigantic Irish Elks have been placed on the Coal Measure!!! So that this carefully constructed and truthful section for visual teaching… has been reversed, stultified or rendered mischievous by the supposed guardians of the Crystal Palace property.
>
> HAWKINS, 1875
> (QUOTED IN BRIERLEY *ET AL.*, 2018, PP. 3–4)

In the same letter Hawkins offered to repair the damaged components without charge and using 'the severest economy in the use of the necessary materials that might be required', but the Crystal Palace Company did not take up his offer. Hawkins' reaction would have been even more exasperated had he known what was to follow in the twentieth century, when – as the park entered ownership of the Ministry of Defence – the palaeontological sculptures were reportedly used for target practice. Later, with the park back in public ownership, a murky period of sculpture loss and damage centred around the Tertiary Island, coinciding with the 1953 building of a petting zoo (The 'Pets Corner'), an attraction expanded in 1960 to a larger zoo. Photographs of this development show a low level of regard for the sculptures: if they could not be moved, they were simply incorporated into animal enclosures (McCarthy and Gilbert, 1994). Thus, for a time, Hawkins' convention-defying *Megatherium* became a climbing frame for goats.

Misfortune followed for the Primary Strata in the 1960s when the development of the Crystal Palace National Sport Centre saw major landworks around the park, including the redesign of the water course adjacent to the Mountain Limestone and Coal Measures. This resulted in the loss of over 50 per cent of the original display (Doyle and Robinson, 1993). Lest it be thought that only building and park development have damaged the sculptures, historic conservation has also been mishandled and resulted in loss of original, authentic park components. The first known round of repair efforts for the Geological Court, overseen by the British Museum of Natural History's William Elgin Swinton (*see* Fig. 13.4), included replacing the lost '*Plesiosaurus*' *macrocephalus* head with that of '*Plesiosaurus*' *hawkinsii* and, potentially from the same works, many mammal sculptures were incorrectly 'restored' with inaccurate replacement parts. Custodial indifference and mismanagement seem to have been the norm for much of the history of the Geological Court, and are probably the single biggest contributor to the loss and damage of Geological Illustrations and palaeontological models.

Fig. 13.4 Conservation works have taken place at various times and scales in the Geological Court. The oldest documented site-wide repair programme was executed in 1959 under direction of Polish sculptor Adam Bienkowski (shown here attending to *Teleosaurus*) in consultation with British Museum (Natural History) palaeontologist William Elgin Swinton. Conservation tasks included replacing lost body parts, sealing cracks and repainting the sculptures.

The modern era of Geological Court conservation

Following decades of relatively callous treatment, the late twentieth century saw the beginnings of our modern approach to preserving the Geological Court. In 1973 the entire site – from the sculptures down to the ground holding their foundations – was granted Grade II heritage status by Historic England, the statutory public body that oversees and celebrates England's historic buildings and environments, reporting to the British Parliament. This protected the entire site from further development and began a more considerate era of conservation, where scientific approaches to testing and documentation ensured the historic materials used at the site received expert treatment. From this point on, conservation work on the Geological Court could only be undertaken by accredited professionals committed to industry standards of material use and documentation.

Such was the approach instigated in the 1990s where, after a series of detailed investigations and consultations, the most extensive and detailed round of conservation conducted on the Geological Court to date was executed from 2001–03. Performed by the Morton Partnership under guidance from Peter Doyle, this not only repaired the degraded paleontological sculptures but also replaced several missing ones (*Anoplotherium* and the Oolite pterosaurs) with new, fibreglass versions crafted by John Warne (Fredrica Banks Sculpture). They also recreated and repaired the Geological Illustrations using authentic stone from historically appropriate quarries across the UK and cleared the entire site of damaging or obscuring plant growth. This brought the Geological Court closer to its original state than at any other point in its recent history.

The Morton Partnership version of the Geological Court is, essentially, still the one we know today. Subsequent conservation works and maintenance have continued but have been selective rather than holistic. Important stabilization and rejuvenation were performed on a number of Secondary Island sculptures by Cliveden Conservation and Skillington Workshop between 2016 and 2017 (*see* Fig. 13.5) but much of the site, at time of writing, has not received conservator attention or even basic surface maintenance for years, leaving some components badly degraded, neglected and overgrown (*see* Fig. 13.3).

It is in this context that the future of the Geological Court is being discussed, and several important developments around the Crystal Palace Dinosaurs are setting the tone for future preservation work. In 2007 the heritage classification

Fig. 13.5 Conservation on the enaliosaurs by Skillington Workshop, 2016–17. **A)** Doff cleaning (a type of steam cleaning) of *Ichthyosaurus communis*; **B–C)** interior details of *I. communis*, showing brickwork (B) and root damage through mortar and brickwork of spine (C); **D)** raking out damaged material in the head of *Leptonectes*; **E)** stripping of '*Plesiosaurus' macrocephalus* to reveal unexpected details of iron and timber neck supports (indicated by red arrows); **F)** mortar filling on *I. communis* before final surface finish; **G)** reinstatement of all possible original material; **H)** painting, the final stage.

of the site was elevated to Grade I by Historic England, the highest mark of recognition in the UK and shared by monuments like Stonehenge, St Paul's Cathedral and the Houses of Parliament. They were further placed on the 'Heritage At Risk Register' in 2020 to escalate their plight to a level of national recognition and urgent priority. Alongside this, the founding of the Friends of Crystal Palace Dinosaurs in 2013 introduced a dedicated body of champions and researchers to monitor, advise, fundraise and engage the public about the significance of the Geological Court. Working with the owners of the site, Bromley Council, and the new custodians, the Crystal Palace Park Trust, four organizations are now focused on the long-term preservation of the Crystal Palace Dinosaurs.

Underpinning the many tasks of this team are the questions about what components of the Geological Court should be conserved, and how they are to be maintained. Is the value of the Geological Court found in the retention of original components as crumbling but physically authentic Victorian relics, or in retaining some degree of completeness, even if that means elements are replaced with modern replicas? By and large, the former approach is followed, in concordance with conservation ethics. The Geological Court is listed on the historic register as a complete site, with its value based around retaining its original, integrated narrative. Thus, while every effort is made to conserve original material, the manufacture of replacement elements is also seen as necessary, as is the returning of lost elements to their original locations when feasible.

There are echoes of the original Crystal Palace Company's philosophy in this approach: just as they prioritized recreated history over real or replica artefacts, the Crystal Palace Dinosaurs are not allowed to become 'historic artefacts' themselves. Their highest cultural value is perceived as retaining the original appearance and layout of the Geological Court, even if that means modern reconstructions in the guise of re-sculpted concrete body parts, fibreglass models, and 3D-printed prostheses must be bolted onto or slotted alongside Victorian originals. This push for relative authenticity, of course, must be weighed against the modern demands of the site and its security. A 2021 addition to the site, a swing bridge at the southwest point of the Secondary Island, is an example of such necessity. Here, historical authenticity is sacrificed for permanent but controlled access to the Secondary Island to enable routine monitoring, maintenance and conservation work, without which the site will not survive.

Fig. 13.6 The conservation value of 3D imaging is demonstrated by repairs to the 2020 damage and prosthetic intervention on *Megalosaurus*. **A)** 3D image of broken surface; **B)** 3D scan before damage, taken by Rhys Griffin and Anthony Lewis, FCPD, providing a fine-detail 3D archival record; **C–E)** individually scanned, colour-coded and digitally reconstructed recovered fragments, showing that essentially all original material was retrieved and could fit onto the broken surface, and informing the shape of the prosthesis; **F)** the broken *Megalosaurus* with new internal framework for prosthesis; **G–H)** installation of 3D printed plastic prosthesis by a team from Taylor Pearce Conservation. *See* Fig. 1.13 for the finished product.

Fig. 13.7 The first significant addition to the Geological Court in decades was the 2021 installation of a swing bridge at the southwestern end of the Secondary Island. The previous permanent, gated means of approach was removed in 2016–17 during repairs to the weir and without reliable, secured access, the core of this Grade I site could not be accessed. The new bridge was crowdfunded through public, civic and business donations to the Friends of Crystal Palace Dinosaurs, designed by Tonkin Liu, engineered by Arup and fabricated by Cake Industries.

Chapter 13 – Past becomes future: the conservation of the Geological Court

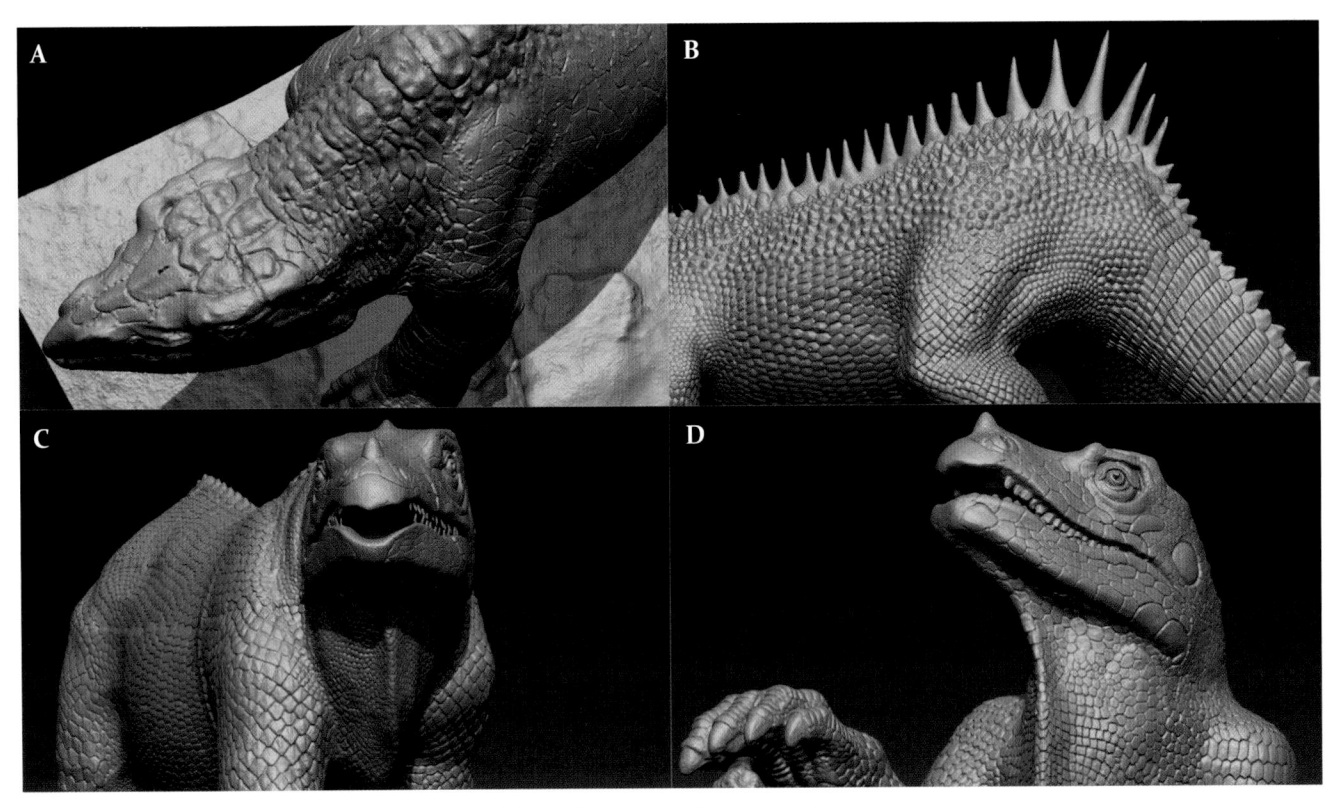

Fig. 13.8 The future of image capture at the Geological Court: digital renders (scans) made by photogrammetry (hundreds or thousands of photographs digitally stitched together into a 3D model). Such imagery can be used for object archiving, material and structural stress mapping, and identifying areas vulnerable to water damage, insolation and vegetation, among other functions. Scanned portions of **A)** *Megalosaurus*; **B)** *Hylaeosaurus*, **C)** the standing *Iguanodon* and **D)** resting *Iguanodon* are shown here. Greyscale rendering such as this emphasizes subtleties in muscle and scale definition, but original surface colours can be included in scan data as well.

This, in turn, underscores the importance of understanding the original layout, intentions and guise of the Geological Court. All attempts to restore the site are, in reality, based on the best contemporary understanding of its original construction, and this continues to evolve as new interpretations and data arise. To that end, an important goal of modern conservators – and a specific objective for the Friends of Crystal Palace Dinosaurs – involves analysis and reinterpretation of the site and related archives to better inform conservation actions. New technologies allow for these approaches to transcend previous efforts at documenting the site. 3D digital imaging, for example, provides unprecedented visual records of the sculptures that can inform the construction of prostheses as well as, through comparing scans, modelling of structural failure (Carpenter, 2017; *see* Fig. 13.8). Uncovering the history and details of the Court in this way provides material for another important conservation mission: dissemination of these findings through public outreach to re-energize interest in the story and plight of this familiar, if somewhat forgotten, corner of south London.

Looking after the Crystal Palace Dinosaurs is, for obvious reasons, mostly a job performed by specialists and experts dedicating professional experience to the site. This is not to say that assisting the conservation and understanding of the Geological Court is an exclusive club, however. To the contrary, there is much we can all do to help. Historic information about the site, especially from before the twenty-first century, is especially sought after and – as is extensively documented in this book – is often in short supply.

There is a wealth of information, however, in old family photographs and videos taken at the Geological Court, as well as in half-forgotten promotional leaflets, obscure books and even personal anecdotes about visiting or working in the park. All may be useful in forming a more complete picture of the history of the site, and readers are encouraged, if they

Fig. 13.9 The Secondary Island dinosaurs, *Teleosaurus* and enaliosaurs in December 2017, photographed shortly after conservation works on the marine reptiles. The difference between sculptures that have received restoration attention and those that have not is evident, highlighting the need for regular conservation and maintenance across the entire Geological Court. As the site approaches its 170th year, the development of such a scheme is essential to its long-term survival.

have unusual information or documentation of this kind, to contact the Friends of Crystal Palace Dinosaurs. Readers can also follow and support the conservation and outreach efforts of the Friends via their website (cpdinosaurs.org) and social media channels. A minor industry of Crystal Palace Dinosaur products now exists among independent artists and merchants, many of whom donate money to the Friends with every purchase. A final, and perhaps the most rewarding, means of assisting the Crystal Palace Dinosaurs is to visit them directly. This not only provides an experience of the Geological Court that no photograph or video can substitute, but a clear show of public interest has obvious value in promoting their maintenance needs to fundraisers and heritage charities.

There is, of course, little question as to why the Crystal Palace Dinosaurs deserve the continued consideration of conservators. The Geological Court is a unique, historically important site with a cultural, scientific and artistic worth that vastly transcends any monetary value we could assign to it. Its importance can only be measured in the ambition it represents from Paxton, Ansted, Hawkins, Campbell and Owen; in the advancements it provided to palaeoartistry and public education; as well as its many peculiarities and Victorian eccentricities. We are fortunate that this last remaining exhibit from the magnificent, if imperfect, Crystal Palace project still exists to draw our wonder and curiosity, and still provides a window into a vision of Deep Time that would otherwise be confined to history books. The true legacy of the Geological Court is not as the decaying remnant of a failed pleasure park or as a series of inaccurate dinosaur models, but as a collective responsibility that successive generations are privileged to look after, study and reinterpret in their own light, all the while continuing the mission of inspiration and education infused into its rocky landscapes, iron skeletons and concrete flesh by those who created it.

References

Abel, O. (1925). *Geschichte und Methode der Rekonstruktion vorzeitlicher Wirbeltiere*. G. Fischer.

Anonymous. (1854). The Geology of the Crystal Palace. Hogg's Instructor, 2, 279–286.

Anonymous. (1871). *Crystal Palace: guide to the Palace and Park*. R.K. Burt.

Ansted, D. T. (1856). *Geological Science: including the practice of geology and the elements of physical geography. With a copious glossarial index*. Houlston & Stoneman.

Argot, C. (2008). Changing views in paleontology: the story of a giant (*Megatherium*, Xenarthra). In: Sargis, E. J., Dagosto, M. (eds). *Mammalian Evolutionary Morphology: A Tribute to Frederick S. Szalay*. Springer, 37–50 pp.

Bajdek, P., Qvarnström, M., Owocki, K., Sulej, T., Sennikov, A. G., Golubev, V. K., & Niedźwiedzki, G. (2016). Microbiota and food residues including possible evidence of pre-mammalian hair in Upper Permian coprolites from Russia. Lethaia, 49(4), 455–477.

Bargo, M. S., Toledo, N., & Vizcaíno, S. F. (2006). Muzzle of South American Pleistocene ground sloths (Xenarthra, Tardigrada). Journal of Morphology, 267(2), 248–263.

Barrett, P. M. & Maidment, S. C. R. (2011). Armoured dinosaurs. In: Batten, D. J. (ed.) *English Wealden fossils*. The Palaeontological Association (London), 391–406 pp.

Beaver, P. (1986). *The Crystal Palace: a portrait of Victorian enterprise*. Phillimore & Company.

Becker, H. (1911). *Alte und neue Rekonstruktionen ausgestorbener Tiere*. Die Umschau.

Benson, R. B. (2010). A description of *Megalosaurus bucklandii* (Dinosauria: Theropoda) from the Bathonian of the UK and the relationships of Middle Jurassic theropods. Zoological Journal of the Linnean Society, 158(4), 882–935.

Benson, R. B., Bates, K. T., Johnson, M. R., & Withers, P. J. (2011). Cranial anatomy of *Thalassiodracon hawkinsii* (Reptilia, Plesiosauria) from the Early Jurassic of Somerset, United Kingdom. Journal of Vertebrate Paleontology, 31(3), 562–574.

Benton, M. J., & Gower, D. J. (1997). Richard Owen's giant Triassic frogs: archosaurs from the Middle Triassic of England. Journal of Vertebrate Paleontology, 17(1), 74–88.

Blows, W. T. (2015). *British polacanthid dinosaurs: observations on the history and palaeontology of the UK polacanthid armoured dinosaurs and their relatives*. Siri Scientific Press.

Bowden, A. J., Tresise, G. R., & Simkiss, W. (2010). *Chirotherium*, the Liverpool footprint hunters and their interpretation of the Middle Trias environment. In: Moody, R. T. J., Buffetaut, E., Naish, D. & Martill, D. M. (eds). *Dinosaurs and other extinct saurians: a historical perspective*. Geological Society Special Publications, 343(1), 209–228.

Bramwell, V., & Peck, R. M. (2008). *All in the bones: a biography of Benjamin Waterhouse Hawkins*. Academy of Natural Sciences.

Brierley, L., Michel, E., Lewis, A., Aldhous, C., & Olmstead, L. (2018). How the 'Seven Deadly Agents of Destruction' can help preserve the Crystal Palace Dinosaurs. In: Sunara, S. and Thorn, A.(eds). *The Conservation of Sculpture Parks*. Archetype Publications, pp. 164–176.

Buckland, W. (1824). Notice on the *Megalosaurus* or great Fossil Lizard of Stonesfield. 1(2), 390–396.

Buckland, W. (1836). *The Bridgewater Treatises on the power wisdom and goodness of god as manifested in the creation: geology and mineralogy considered with reference to natural theology, Treatise VI*, (2 vols). William Pickering.

Cadbury, D. (2000). *The dinosaur hunters: a story of scientific rivalry and the discovery of the prehistoric world*. Fourth Estate.

Cain, J. (2014). George Baxter's famous print of Crystal Palace Dinosaurs re-examined. Profjoecain.net blog post. https://profjoecain.net/baxter-famous-print-crystal-palace-dinosaurs, retrieved 18/08/2021.

Campos, D., & Kellner, A. W. A. (1985). Panorama of the flying reptiles study in Brazil and South America. Anais da Academia Brasileira de Ciências, 57(4), 453–466.

Carpenter, India, 2017. Development of a new methodology for mapping and monitoring the condition of 3dimensional cultural heritage: Case study – the standing *Iguanodon* in Crystal Palace Park using photogrammetry and 3D digitalisation to record condition and to map and monitor deterioration. Final Year Research Paper PG Diploma in Conservation Department, City & Guilds of London Art School. 130pp.

Chudinov, P.K. (1968). Structure of the integuments of theromorphs. Doklady Akademii Nauk SSSR, 179(1), 226–229.

Conway, J., Kosemen, C. M., & Naish, D. (2013). *All yesterdays: unique and speculative views of dinosaurs and other prehistoric animals*. Irregular Books.

Cope, E. D. (1867). The fossil reptiles of New Jersey. The American Naturalist, 1(1), 23–30.

Cope, E. D. (1869). The fossil reptiles of New Jersey (Continued). The American Naturalist, 3(2), 84–91.

Craddock, J. P. (2016). *People of the Palace: the Crystal Palace Company & Trust*. Crystal Palace Foundation.

Cuvier, G. (1812). *Recherches sur les ossemens fossiles de quadrupèdes: où l'on rétablit les caractères de plusieurs espèces d'animaux que les révolutions du globe paroissent avoir détruites, Volume 4*. Deterville.

Cuvier, G. (1822). *Recherches sur les Ossemens Fossiles (Nouvelle Edition), Volume 2, Part 1*. Chez G. Dufour et E. D'Ocagne.

Cuvier, G. (1827). *Essay on the theory of the Earth*. W. Blackwood.

Damiani, R. J. (2001). A systematic revision and phylogenetic analysis of Triassic mastodonsauroids (Temnospondyli: Stereospondyli). Zoological Journal of the Linnean Society, 133(4), 379–482.

Davidson, J. P. (2005). Henry A. Ward, catalogue of casts of fossils (1866) and the artistic influence of Benjamin Waterhouse Hawkins on Ward. Transactions of the Kansas Academy of Science, 108(3), 138–148.

Dawson, G. (2016). *Show me the bone: reconstructing prehistoric monsters in nineteenth-century Britain and America*. University of Chicago Press.

Dean, D. R. (1999). *Gideon Mantell and the discovery of dinosaurs*. Cambridge University Press.

Desmond, A. J. (1976). *The hot-blooded dinosaurs: a revolution in palaeontology*. Dial Press.

Dodson, P. (1998). *The horned dinosaurs: a natural history*. Princeton University Press.

Doyle, P. (2008). A vision of 'deep time': the 'Geological Illustrations' of Crystal Palace Park, London. In: Burek, C. V. & Prosser, C. D. (eds). *The History of Geoconservation*. Geological Society Special Publications, 300(1), 197–205.

Doyle, P., & Robinson, E. (1993). The Victorian 'Geological Illustrations' of Crystal Palace Park. Proceedings of the Geologists' Association, 104(3), 181–194.

Doyle, P., & Robinson, E. (1995). Report of a field meeting to Crystal Palace Park and West Norwood Cemetery, 11 December, 1993. Proceedings of the Geologists' Association, 106(1), 71–78.

Elgin, R. A., Hone, D. W., & Frey, E. (2011). The extent of the pterosaur flight membrane. Acta Palaeontologica Polonica, 56(1), 99–111.

Emling, S. (2009). *The fossil hunter: dinosaurs, evolution, and the woman whose discoveries changed the world*. St. Martin's Press.

Fariña, R. A. (2002). *Megatherium*, the hairless: appearance of the great Quaternary sloths (Mammalia; Xenarthra). Ameghiniana, 39(2), 241–244.

Fariña, R. A., Vizcaíno, S. F., & De Iuliis, G. (2013). *Megafauna: giant beasts of pleistocene South America*. Indiana University Press.

Figuier, L. (1864). *La Terre avant le déluge*. L. Hachette.

Flammarion, C. (1886). *Le Monde avant la création de l'homme: origines de la terre, origines de la vie, origines de l'humanité*. C. Marpon et E. Flammarion.

Foth, C., Tischlinger, H., & Rauhut, O. W. (2014). New specimen of *Archaeopteryx* provides insights into the evolution of pennaceous feathers. Nature, 511(7507), 79–82.

Frey, E., Mulder, E. W., Stinnesbeck, W., Rivera-Sylva, H. E., Padilla-Gutiérrez, J. M., & González-González, A. H. (2017). A new polycotylid plesiosaur with extensive soft tissue preservation from the early Late Cretaceous of northeast Mexico. Boletín de la Sociedad Geológica Mexicana, 69(1), 87–134.

Gardiner, B. G. (1991). Clift, Darwin, Owen and the Dinosauria... (2). The Linnean. 7(1), 8–14.

Geist, V. (1999). *Deer of the world*. Swan Hill Press, Shrewsbury.

Goldfuss, A. (1831). Beiträge zur Kenntnis verschiedener Reptilien der Vorwelt. Nova Acta Physico-Medica Academiae Caesareae Leopoldino-Carolinae Naturae Curiosorum, 15, 61–128.

Goldfuss, A. (1845). Der Schädelbau des *Mosasaurus*, durch Beschreibung einer neuen Art dieser Gattung erläutert. Nova Acta Academa Ceasar Leopoldino-Carolinae Germanicae Natura Curiosorum, 21, 1–28.

Guthrie, R. D. (2005). *The nature of Paleolithic art*. The University of Chicago Press.

Hallett, M. (1987). The scientific approach to the art of bringing dinosaurs to life. In: Czerkas, S. J. & Olson, E. C. (eds). *Dinosaurs past and present vol. I*. Natural History Museum of Los Angeles County/University of Washington Press, 96–113 pp.

Hawkins, B. W. (1854). On visual education as applied to geology, illustrated by diagrams and models of the geological restorations at the Crystal Palace. Journal of the Society of Arts, 78, 443–449.

Hawkins, B. W. (1876). *The artistic anatomy of the dog and deer (vol. 20)*. Winsor and Newton.

Hawkins, T. (1834). *Memoirs of Ichthyosauri and Plesiosauri, extinct monsters of the ancient Earth: with twenty-eight plates, copies from specimens in the author's collection of fossil organic remains*. Relfe and Fletcher.

Hirst Conservation. (2000). 4844 – Historic paint research. Crystal Palace Dinosaurs. [Unpublished report]

Hooker, J. J. (2007). Bipedal browsing adaptations of the unusual Late Eocene–earliest Oligocene tylopod *Anoplotherium* (Artiodactyla, Mammalia). Zoological Journal of the Linnean Society, 151, 609–659.

Howgate, M. (2015). The Wisbech Museum Waterhouse Hawkins prehistoric models: were these the first ever prehistoric models? Prehistoric Times, 114, 28–29.

Hutchinson, N. H. (1893). *Creatures of other days*. Chapman & Hall.

Hutchinson, N. H. (1894). *Extinct monsters: a popular account of some of the larger forms of ancient animal life*. Chapman & Hall.

Huxley, T. H. (1859). On *Rhamphorhynchus bucklandi*, a pterosaurian from the Stonesfield Slate. Quarterly Journal of the Geological Society, 15(1–2), 658–670.

Jäger, K. R., Tischlinger, H., Oleschinski, G., & Sander, P. M. (2018). Goldfuß was right: soft part preservation in the Late Jurassic pterosaur *Scaphognathus crassirostris* revealed by reflectance transformation imaging (RTI) and UV light and the auspicious beginnings of paleo-art. Palaeontologia Electronica, 21(3), 1–20.

Johnson, M. M., Young, M. T., & Brusatte, S. L. (2020). The phylogenetics of Teleosauroidea (Crocodylomorpha, Thalattosuchia) and implications for their ecology and evolution. PeerJ, 8, e9808.

Kammerer, C. F., Angielczyk, K. D., & Fröbisch, J. (2011). A comprehensive taxonomic revision of *Dicynodon* (Therapsida, Anomodontia) and its implications for dicynodont phylogeny, biogeography, and biostratigraphy. Journal of Vertebrate Paleontology, 31(sup1), 1–158.

King, G. M. (1981). The functional anatomy of a Permian dicynodont. Philosophical Transactions of the Royal Society of London. B, Biological Sciences, 291(1050), 243–322.

Knight, N. (2001). *The Empire on display: ethnographic exhibition and the conceptualization of human diversity in post-emancipation Russia*. The National Council for Eurasian and East European Research.

Knoll, F., & López-Antoñanzas, R. (2011). The true world's first sculptures of antediluvian animals, which never were.... Comptes Rendus Palevol, 10(7), 597–604.

Leith, I. (2005). *DeLamotte's Crystal Palace: a Victorian pleasure dome revealed*. Historic England Press.

Lescaze, Z. (2017). *Paleoart: visions of the prehistoric past*. Taschen.

Lindgren, J., Everhart, M. J., & Caldwell, M. W. (2011). Three-dimensionally preserved integument reveals hydrodynamic adaptations in the extinct marine lizard *Ectenosaurus* (Reptilia, Mosasauridae). PLoS One, 6(11), e27343.

Lindgren, J., Kaddumi, H. F., & Polcyn, M. J. (2013). Soft tissue preservation in a fossil marine lizard with a bilobed tail fin. Nature Communications, 4, 2423.

Lindgren, J., Sjövall, P., Thiel, V., Zheng, W., Ito, S., Wakamatsu, K., ... & Eriksson, M. E. (2018). Soft-tissue evidence for homeothermy and crypsis in a Jurassic ichthyosaur. Nature, 564(7736), 359.

Lingham-Soliar, T. (1995). Anatomy and functional morphology of the largest marine reptile known, *Mosasaurus hoffmanni* (Mosasauridae, Reptilia) from the Upper Cretaceous, Upper Maastrichtian of the Netherlands. Philosophical Transactions of the Royal Society of London. Series B: Biological Sciences, 347(1320), 155–180.

Lister, A. M. (1987). *Megaceros* or *Megaloceros*? The nomenclature of the Giant Deer. Quaternary Newsletter, 52, 14–16.

Lister, A. M. (1994). The evolution of the Giant Deer, *Megaloceros giganteus* (Blumenbach). Zoological Journal of the Linnean Society, 112(1–2), 65–100.

Liston, J. J. (2010). 2000 AD and the new 'Flesh': first to report the dinosaur renaissance in 'moving' pictures. In: Moody, R. T. J., Buffetaut, E., Naish, D. & Martill, D. M. (eds). *Dinosaurs and other extinct saurians: a historical perspective*. Geological Society Special Publications, 343(1), 335–360.

Liu, S., Smith, A. S., Gu, Y., Tan, J., Liu, C. K., & Turk, G. (2015). Computer simulations imply forelimb-dominated underwater flight in plesiosaurs. PLoS Computational Biology, 11(12), e1004605.

Lomax, D. R., & Tamura, N. (2014). *Dinosaurs of the British Isles*. Siri Scientific Press.

Lomax, D. R., Larkin, N. R., Boomer, I., Dey, S., & Copestake, P. (2019). The first known neonate *Ichthyosaurus communis* skeleton: a rediscovered specimen from the Lower Jurassic, UK. Historical Biology, 31(5), 600–609.

Lyell, C. (1855). *A manual of elementary geology: or, the ancient changes of the earth and its inhabitants as illustrated by geological monuments*. J. Murray.

Mantell, G. A. (1825). Notice on the *Iguanodon*, a newly discovered fossil reptile, from the sandstone of Tilgate, in Sussex. Philosophical Transactions of the Royal Society of London, (115), 179-186.

Mantell, G. A. (1827). *Illustrations of the geology of Sussex: with figures and descriptions of the Fossils of Tilgate*. Lupton Relfe.

Mantell, G. A. (1833). *The geology of the south-east of England*. Longman, Rees, Orme, Brown, Green and Longman.

Mantell, G. A. (1839). *The wonders of geology; of a familiar exposition of geological phenomena; being the substance of a course of lectures delivered at Brighton, 3rd edition. (2 vols)*. Relfe & Fletcher.

Mantell, G. A. (1848). On the structure of the jaws and teeth of the *Iguanodon*. Philosophical Transactions of the Royal Society of London, (138), 183–202.

Mantell, G. A. (1849) Additional observations on the osteology of the *Iguanodon* and *Hylaeosaurus*. With notes on the vertebral column of the *Iguanodon*; by A. G. Melville. Philosophical Transactions of the Royal Society of London, CXXXIX, 271–305.

Mantell, G. A. (1851). *Petrifactions and their teachings: or, a handbook to the gallery of organic remains of the British Museum (Vol. 6)*. HG Bohn.

Marsh, O. C. (1871), Note on a new and gigantic species of pterodactyle. American Journal of Science, 1(3), 472.

[Martineau, H.]. (1854). The Crystal Palace. Westminster Review, 62, 534–50.

McCarthy, S., & Gilbert, M. (1994). *The Crystal Palace Dinosaurs: the story of the world's first prehistoric sculptures*. Crystal Palace Foundation.

McDermott, E. (1854). *Routledge's guide to the Crystal Palace and Park at Sydenham*. Routledge, London.

McGowan, C. (1996). Giant ichthyosaurs of the Early Jurassic. Canadian Journal of Earth Sciences, 33(7), 1011–1021.

McGowan, C., & Motani, R. (2003). *Handbook of Paleoherpetology, Part 8 Ichthyopterygia*. Verlag Dr. Friedrich Pfeil.

McHenry, C. R., Cook, A. G., & Wroe, S. (2005). Bottom-feeding plesiosaurs. Science, 310(5745), 75–75.

Melchor, R. N., Perez, M., Cardonatto, M. C., & Umazano, A. M. (2015). Late Miocene ground sloth footprints and their paleoenvironment: *Megatherichnum oportoi* revisited. Palaeogeography, Palaeoclimatology, Palaeoecology, 439, 126–143.

Milner, R. (2012). *Charles R. Knight: The artist who saw through time*. Abrams.

Moser, M., & Schoch, R. (2007). Revision of the type material and nomenclature of *Mastodonsaurus giganteus* (Jaeger) (Temnospondyli) from the Middle Triassic of Germany. Palaeontology, 50(5), 1245–1266.

Muscutt, L. E., Dyke, G., Weymouth, G. D., Naish, D., Palmer, C., & Ganapathisubramani, B. (2017). The four-flipper swimming method of plesiosaurs enabled efficient and effective locomotion. Proceedings of the Royal Society B: Biological Sciences, 284(1861), 20170951.

Naish, D. (2010). Pneumaticity, the early years: Wealden Supergroup dinosaurs and the hypothesis of saurischian pneumaticity. In: Moody, R. T. J., Buffetaut, E., Naish, D. & Martill, D. M. (eds). *Dinosaurs and other extinct saurians: a historical perspective*. Geological Society Special Publications, 343(1), 229–236.

Nash, J., Haghe, L., & Roberts, D. (1854). Dickinson's comprehensive pictures of the Great Exhibition of 1851. Dickinson.

Nieuwland, I. (2019). *American dinosaur abroad: a cultural history of Carnegie's Plaster Diplodocus*. University of Pittsburgh Press.

Noè, L. F., Taylor, M. A., & Gómez-Pérez, M. (2017). An integrated approach to understanding the role of the long neck in plesiosaurs. Acta Palaeontologica Polonica, 62(1), 137–162.

Norman, D. B. (1991). *Dinosaur!* Boxtree.

Norman, D. B. (1993). Gideon Mantell's 'Mantel-piece': the earliest well-preserved ornithischian dinosaur. Modern Geology, 18(2), 225–245.

Norman, D. B. (2000). Professor Richard Owen and the important but neglected dinosaur *Scelidosaurus harrisonii*. Historical Biology, 14(4), 235–253.

Norman, D. B. (2011). Ornithopod dinosaurs. In: Batten, D. J. (ed). *English Wealden fossils*. The Palaeontological Association (London). 407–475 pp.

Norman, D. B. (2013). On the taxonomy and diversity of Wealden iguanodontian dinosaurs (Ornithischia: Ornithopoda). Revue de Paléobiologie, 32(2), 385–404.

Norman, D. B. (2015). On the history, osteology, and systematic position of the Wealden (Hastings group) dinosaur *Hypselospinus fittoni* (Iguanodontia: Styracosterna). Zoological Journal of the Linnean Society, 173(1), 92–189.

O'Keefe, F. R., & Chiappe, L. M. (2011). Viviparity and K-selected life history in a Mesozoic marine plesiosaur (Reptilia, Sauropterygia). Science, 333(6044), 870–873.

O'Sullivan, M., & Martill, D. (2018). Pterosauria of the Great Oolite Group (Middle Jurassic, Bathonian) of Oxfordshire and Gloucestershire, England. Acta Palaeontologica Polonica, 63(4), 617–644.

Ortega, F., Escaso, F., & Sanz, J. L. (2010). A bizarre, humped Carcharodontosauria (Theropoda) from the Lower Cretaceous of Spain. Nature, 467(7312), 203–206.

Owen, R. (1840a). Exhibition of a bone of an unknown struthious bird from New Zealand. Proceedings of the Zoological Society, 7, 169–71.

Owen, R. (1840b). Note on the dislocation of the tail at a certain point observable in the skeleton of many Ichthyosauri. Transactions of the Geological Society of London, 2(3), 511–514.

Owen, R. (1840c). A description of a specimen of the *Plesiosaurus Macrocephalus*, Conybeare, in the collection of Viscount Cole, MP, DCL, FGS, &c. Transactions of the Geological Society of London, 2(3), 515–535.

Owen, R. (1841). A description of some of the soft parts, with the integument, of the hind-fin of the *Ichthyosaurus*, indicating the shape of the fin when recent. Transactions of the Geological Society of London, 2(1), 199–201.

Owen, R. (1842a). Notice of a fragment of the femur of a gigantic bird in New Zealand. Transactions of the Zoological Society, 3, 29–32.

Owen, R. (1842b). Description of the skeleton of an extinct gigantic sloth: *Mylodon Robustus*, Owen, with observations on the osteology, natural affinities, and probable habits of the megatherioid quadrupeds. J. van Voorst.

Owen, R. (1842c). Report on British fossil reptiles, part II. Report for the British Association for the Advancement of Science, Plymouth, 1841, 60–204.

Owen, R. (1842d). On the teeth of species of the genus *Labyrinthodon* (*Mastodonsaurus* of Jaeger) common to the German Keuper formation and the lower sandstone of Warwick and Leamington. Transactions of the Geological Society of London, 2(2), 503–513.

Owen, R. (1842e). Description of parts of the skeleton and teeth of five species of the genus *Labyrinthodon* (*Lab. leptognathus*, *Lab. pachygnathus*, and *Lab. ventricosus*, from the Coton-end and Cubbington Quarries of the Lower Warwick Sandstone; *Lab. Jægeri*, from Guy's Cliff, Warwick; and *Lab. scutulatus*, from Leamington); with remarks on the probable identity of the *Cheirotherium* with this genus of extinct batrachians. Transactions of the geological Society of London, 2(2), 515–543.

Owen, R. (1845). Report on the reptilian fossils of South Africa: part I.—Description of certain fossil crania, discovered by AG Bain, Esq., in sandstone rocks at the south-eastern extremity of Africa, referable to different species of an extinct genus of reptilia (*Dicynodon*), and indicative of a new tribe or sub-order of Sauria. Transactions of the Geological Society of London, 2(1), 59–84.

Owen, R. (1846). *A history of British fossil mammals, and birds*. J. Van Voorst.

Owen, R. (1851) Monograph on the fossil Reptilia of the Cretaceous formations. Palaeontographical Society Monograph, 5, 1–118.

Owen, R. (1854). *Geology and inhabitants of the ancient world*. Crystal Palace Library.

Owen, R. (1855). The fossil Reptilia of the Wealden and Purbeck formations. Part II: Dinosauria (*Iguanodon*). Palaeontographical Society Monograph, 1854, 1–54.

Owen, R. (1856). The fossil Reptilia of the Wealden formations. Part III, *Megalosaurus bucklandii*. Palaeontographical Society Monographs, 9, 1–26.

Owen, R. (1857). Description of the lower jaw and teeth of an anoplotherioid quadruped (*Dichobune ovina*, Ow.) of the size of the *Xiphodon gracilis*, Cuv., from the Upper Eocene Marl, Isle of Wight. Quarterly Journal of the Geological Society, 13, 254–260.

Owen, R. (1858). Monograph on the fossil Reptilia of the Wealden and Purbeck formations. Part IV. Dinosauria (*Hylaeosaurus*). Palaeontographical Society Monograph, 10, 1–26.

Owen, R. (1863). Monographs on the British Fossil Reptilia from the Oolitic Formations. Part Second, containing *Scelidosaurus harrisonii* and *Pliosaurus grandis*. Palaeontographical Society Monograph, 14, 1–26.

Owen, R. (1884). *A history of British fossil reptiles*. Cassell.

Owen, R. (1894). *The life of Richard Owen*. J. Murray.

Padian, K. (1997). The rehabilitation of Sir Richard Owen. BioScience, 47(7), 446–453.

Paul, G. S. (2000). A quick history of dinosaur art. In: Paul, G. S. (ed). *The Scientific American Book of Dinosaurs*. Byron Preiss Visual Productions. 107–112 pp.

Pearce, J. C. (1846). Notice of what appears to be the embryo of an *Ichthyosaurus* in the pelvic cavity of *Ichthyosaurus* (*communis*?). Journal of Natural History, 17(109), 44–46.

Pearson, H. S. (1924). A dicynodont reptile reconstructed. Proceedings of the Zoological Society of London, 94(3), 827–855.

Pereda-Suberbiola, J. (1993). *Hylaeosaurus, Polacanthus*, and the systematics and stratigraphy of Wealden armoured dinosaurs. Geological Magazine, 130(6), 767–781.

Phillips, S. (1854). *Guide to the Crystal Palace and Park.* Crystal Palace Library.

Phillips, S. (1855). *Guide to the Crystal Palace and Park.* Crystal Palace Library.

Phillips, S. (1856). *Guide to the Crystal Palace and Park.* Crystal Palace Library.

Pinegar, R. T., Loewen, M. A., Cloward, K. C., Hunter, R. J. and Weege, C. J. (2003). A juvenile allosaur with preserved integument from the basal Morrison Formation of Central Wyoming. Journal of Vertebrate Paleontology, 23, 87A–88A.

Plieninger, T. (1844). Beiträge von Theodor Plieninger. 51–132, 12 pls. In Meyer, H. von and Plieninger, T. (eds) *Beiträge zur Paläontologie Württemberg's, enthaltend die fossilen Wirbelthierreste aus den Triasgebilden mit besonderer Rücksicht auf die Labyrinthodonten des Keupers.* E. Schweizerbart. 132 pp.

Prothero, D. R. (2016). *The Princeton field guide to prehistoric mammals.* Princeton University Press.

Prothero, D. R., & Foss, S. E. (eds). (2007). *The evolution of artiodactyls.* John Hopkins University Press.

Rémy, J. A. (1992). Observations sur l'anatomie crânienne du genre *Palaeotherium* (Perissodactyla, Mammalia); mise en évidence d'un nouveau sous-genre, *Franzenitherium*. Palaeovertebrata, 21(3–4), 103–224.

Rémy, J. A. (2004). Le genre *Plagiolophus* (Palaeotheriidae, Perissodactyla, Mammalia): révision systématique, morphologie et histologie dentaires, anatomie crânienne, essai d'interprétation fonctionnelle. Palaeovertebrata, 33(1–4), 17–281.

Ross, F. D. & Mayer, G. C. (1983). On the dorsal armor of the Crocodilia. In: Rhodin, A. G. J. & Miyata, K. (eds) *Advances in Herpetology and Evolutionary Biology*. Museum of Comparative Zoology. 306–331 pp.

Ross, R. M., Duggan-Haas, D., & Allmon, W. D. (2013). The posture of *Tyrannosaurus rex*: why do student views lag behind the science? Journal of Geoscience Education, 61(1), 145–160.

Rudwick, M. J. (1992). *Scenes from deep time: early pictorial representations of the prehistoric world.* University of Chicago Press.

Rudwick, M. J. (1997). *Fossil bones and geological catastrophes: new translations & interpretations of the primary texts.* University of Chicago Press.

Rudwick, M. J. (2005). *Bursting the limits of time: the reconstruction of geohistory in the age of revolution.* University of Chicago Press.

Rupke, N. A. (2009). *Richard Owen: biology without Darwin.* University of Chicago Press.

Secord, J. A. (2004). Monsters at the Crystal Palace. In: de Chadarevian, S., & Hopwood, N. (eds) *Models: the third dimension of science.* Stanford University Press. 138–69 pp.

Seeley, H. G. (1888). Researches on the structure, organization, and classification of the fossil reptilia.—V. On associated bones of a small anomodont reptile, *Keirognathus cordylus* (Seeley), showing the relative dimensions of the anterior parts of the skeleton, and structure of the fore-limb and shoulder girdle. Philosophical Transactions of the Royal Society of London.(B.), (179), 487–501.

Sennikov, A. G. (2019). Peculiarities of the structure and locomotor function of the tail in Sauropterygia. Biology Bulletin, 46(7), 751–762.

Smith, A. S. (2013). Morphology of the caudal vertebrae in *Rhomaleosaurus zetlandicus* and a review of the evidence for a tail fin in Plesiosauria. Paludicola, 9(3), 144–158.

Smith, A. S., & Araújo, R. (2017). *Thaumatodracon wiedenrothi*, a morphometrically and stratigraphically intermediate new rhomaleosaurid plesiosaurian from the Lower Jurassic (Sinemurian) of Lyme Regis. Palaeontographica Abteilung A, 308(4–6), 89–125.

Snively, E., & Russell, A. P. (2007). Functional variation of neck muscles and their relation to feeding style in Tyrannosauridae and other large theropod dinosaurs. The Anatomical Record: Advances in Integrative Anatomy and Evolutionary Biology, 290(8), 934–957.

Spindler, F., Lauer, R., Tischlinger, H., & Mäuser, M. (2021). The integument of pelagic crocodylomorphs (Thalattosuchia: Metriorhynchidae). Palaeontologia Electronica, 24(2), a25.

Street, H. P., & Caldwell, M. W. (2017). Rediagnosis and redescription of *Mosasaurus hoffmannii* (Squamata: Mosasauridae) and an assessment of species assigned to the genus *Mosasaurus*. Geological Magazine, 154(3), 521–557.

Taquet, P., & Padian, K. (2004). The earliest known restoration of a pterosaur and the philosophical origins of Cuvier's Ossemens Fossiles. Comptes Rendus Palevol, 3(2), 157–175.

Taylor, M. P. (2010). Sauropod dinosaur research: a historical review. In: Moody, R. T. J., Buffetaut, E., Naish, D. & Martill, D. M. (eds). *Dinosaurs and other extinct saurians: a historical perspective.* Geological Society Special Publications, 343(1), 361–386.

Torrens, H. S. (2012). Politics and Paleontology: Richard Owen and the invention of dinosaurs. In: Brett-Surman, M.K. Holtz, Jr. T. R., & Farlow, J.O. (eds). *The complete dinosaur, second edition.* Indiana University Press. 24–43 pp.

Upchurch, P., Mannion, P.D., & Barrett, P.M. (2011). Sauropod dinosaurs In: D.J. Batten (ed). *English Wealden fossils*. The Palaeontological Association (London). 476–525 pp.

Wall, W. P. (1980). Cranial evidence for a proboscis in *Cadurcodon* and a review of snout structure in the family Amynodontidae (Perissodactyla, Rhinocerotoidea). Journal of Paleontology, 54, 968–977.

Wellnhofer, P. (1991). *The illustrated encyclopedia of pterosaurs.* Salamander Press.

Williston, S. W. (1914). *Water reptiles of the past and present.* University of Chicago Press.

Witton, M. P. (2013). *Pterosaurs: natural history, evolution, anatomy.* Princeton University Press.

Witton, M. P. (2018). *The palaeoartist's handbook: recreating prehistoric animals in art.* The Crowood Press.

Witzmann, F. (2009). Comparative histology of sculptured dermal bones in basal tetrapods, and the implications for the soft tissue dermis. Palaeodiversity, 2(233), e270.

Woodward, A. S., & Moreno, F. P. (1899). On a portion of mammalian skin, named *Neomylodon listai*, from a Cavern near Consuelo Cove, Last Hope Inlet, Patagonia. Proceedings of the Zoological Society of London, 67, 144–156.

Wyatt, M. D. (1854). Views of the Crystal Palace and Park, Sydenham. Day and Son.

Yahaya, A., Olopade, J. O., Kwari, H. D., & Wiam, I. M. (2012). Investigation of the osteometry of the skull of the one-humped camels: Part II: sex dimorphism and geographical variations in adults. Archivio italiano di anatomia ed embriologia, 117(1), 34–44.

Young, M. T., Rabi, M., Bell, M. A., Foffa, D., Steel, L., Sachs, S., & Peyer, K. (2016). Big-headed marine crocodyliforms and why we must be cautious when using extant species as body length proxies for long-extinct relatives. Palaeontologia Electronica, 19(3), 1–14.

Image credits

Unless noted below or in figure captions, all modern photographs and life reconstructions of fossil animals are © Mark P. Witton.

1.1.	Dinosaurs, *Teleosaurus* and plesiosaurs. © *Kevin Ireland*.
1.7.	Northeast view of Secondary Island. © *Kevin Ireland*.
1.10.	*Megaloceros* stag. © *Kevin Ireland*.
1.11.	*Iguanodon*. © *Kevin Ireland*.
2.2B.	David Thomas Ansted. Lithograph by T. H. Maguire, 1850, after himself. *Wellcome Collection. cc-by 4.0.*
2.3.	The Crystal Palace from the Great Exhibition, installed at Sydenham: sculptures of prehistoric creatures in the foreground. Colour Baxter-process print by G. Baxter. *Wellcome Collection. cc-by 4.0.*
2.3.	Needham's *The Crystal Palace and Park.* © *London Picture Library.*
2.6A.	Benjamin Waterhouse Hawkins. *Wellcome Collection. cc-by 4.0.*
2.6B.	Sir Richard Owen. Lithograph by T. H. Maguire, 1850. *Wellcome Collection. cc-by 4.0.*
2.6C.	Portrait of G.A. Mantell by Davey after Sentier. *Wellcome Collection. cc-by 4.0.*
2.6D.	Georges-Léopold-Chrétien-Frédéric-Dagobert, Baron Cuvier. *Wellcome Collection. cc-by 4.0.*
2.8.	Skeleton of a man, with the skeleton of an elephant. Lithograph by B. Waterhouse Hawkins, 1860. *Wellcome Collection. cc-by 4.0.*
2.10.	Crystal Palace dinosaurs. © *Kevin Ireland*.
2.13.	*Hylaeosaurus*. © *Kevin Ireland*.
3.4A.	Hawkins' drawing of stratigraphy of Geological Court. *From the collections of the Library and Archives, Natural History Museum, London, public domain.*
3.5.	*Ichthyosaurus* and '*Plesiosaurus*'. © *Kevin Ireland*.
3.13.	Secondary Island under construction. Photo by Philip Henry Delamotte. © *British Library Board (Tab.442.a.5, 66).*
3.14.	Secondary Island under construction. Photo by Philip Henry Delamotte. © *British Library Board (Tab.442.a.5, 58).*
3.15–16.	Secondary Island under construction. Photo by Philip Henry Delamotte. © *London Picture Library.* These photographs were provided by the London Picture Archive, which is managed by London Metropolitan Archives (LMA) and part of the City of London Corporation. You can browse LMA's set of the extraordinary photographs taken by Delamotte during the construction of the Crystal Palace at Sydenham on the London Picture Archive website, along with many other historical prints, photographs and maps of London. www.londonpicturearchive.org.uk
3.17.	Tertiary Island in 1855. © *Crystal Palace Foundation.*
3.18.	Hawkins sketches. *From the collections of the Library and Archives, Natural History Museum, London, public domain.*
3.19.	Wisbech & Fenland Museum models. *Reproduced by permission of Wisbech & Fenland Museum.*
3.20C.	Workshop image of *Iguanodon* next to scale model from Measom, G. 1855. *London Metropolitan Archive.*
3.24.	Hawkins sketch of *Iguanodon* banquet. *British Library, cc-by 2.0.*
3.25.	New Year's Eve banquet invitation. *From the collections of the Library and Archives, Natural History Museum, London, public domain.*
4.2.	*Megaloceros* 2017. *Neil Cummings, cc by-sa 2.0.*
4.3.	1864 *Megaloceros*. © *Crystal Palace Foundation.*
4.4.	C. Wapiti. *Larry E Smith, cc-by 2.0.*
4.9C.	Linnaeus' two-toed Sloth (*Choloepus didactylus*) at the Buffalo Zoo. *Dave Pape, public domain.*
4.12.	*Palaeotherium minus* and *P. medium*. *Loz Pycock from London, UK, cc by-sa 2.0.*
4.13E.	*Palaeotherium* in 1958. © *Crystal Palace Foundation.*
4.14B.	African elephant in Addo park (South Africa). *Dariusz Jemielniak ('Pundit'), cc by-sa 2.0.*
4.14D.	Schabrackentapir (*Tapirus indicus*), Tierpark Hellabrunn, München. *Rufus46, cc by-sa 3.0.*
4.18.	*Anoplotherium* group. © *David W. E. Hone*.
4.19B.	Dromedary profile. *Bruno /Germany, from Pixabay.*
5.3.	Rock monitor. *Derek Keats, cc-by 2.0.*
6.1.	*Cimoliopterus cuvieri* 2009. *Ben Sutherland, cc-by 2.0.*
6.2A.	Original Oolite pterosaurs. © *Crystal Palace Foundation.*
6.2B–E.	Oolite pterosaurs in situ and storage. © *Joe Cain.*
8.2.	Siamese crocodile. *lonelyshrimp, in public domain.*
9.1.	*Temnodontosaurus* and *Plesiosaurus*. © *David W. E. Hone.*
9.3C.	*Temnodontosaurus*. *Chris Sampson, cc-by 2.0.*
9.9A.	'*Plesiosaurus*' *macrocephalus*. *Tom Page, cc by-sa 2.0*
10.1A.	'*Labyrinthodon salamandroides*'. © *David W. E. Hone.*
10.4A.	Hawkins' sketch of *Labyrinthodon*. *From the collections of the Library and Archives, Natural History Museum, London, public domain.*
10.6.	*Chirotherium* footprint. © *Crystal Palace Foundation.*
11.1.	*Dicynodon*. © *David W. E. Hone.*
11.3A.	*Chelydra*. *Ontley, in public domain.*
12.2.	1:12 sculptures of Crystal Palace Dinosaurs. © *The Trustees of the Natural History Museum, London.*
12.4–6.	Hawkins' posters and development sketches. *From the collections of the Library and Archives, Natural History Museum, London, public domain.*
12.7.	Hawkins' 1855 map. *From the collections of the Library and Archives, Natural History Museum, London, public domain.*
12.8–9.	Hawkins' 1870s oil paintings. *Princeton University Art Museum, public domain.*
12.16.	Secondary Island view. © *David W. E. Hone.*
Chapter 13	header, *Iguanodon* teeth. © *James Balston Photography.*
13.1A.	*Megaloceros* in snow. *Nikki Dawson, Pixabay.*
13.1B.	Jurassic reptiles in snow. © *Joe Cain.*
13.2F–G.	Cracked *Iguanodon*. © *James Balston Photography.*
13.4.	1950s conservation. © *Crystal Palace Foundation.*
13.5.	Skillington workshop conservation photos. © *Skillington Workshop.*
13.6A–E.	*Megalosaurus* digital scans. © *Rhys Griffin & Anthony Lewis, FCPD.*
13.6F–H.	Installation of *Megalosaurus* prosthesis. *Photos by Chris Redgrave,* © *Historic England Archive.* The Historic England Archive is a gateway to discover more about England's archaeology, historic buildings and social history. It holds over 12 million photographs, drawings, reports and publications from the 1850s to the present day. https://historicengland.org.uk/images-books/photos/
13.7.	Swing bridge © *James Balston Photography.*
13.8.	Dinosaur digital scans. © *Rhys Griffin & Anthony Lewis, FCPD.*
13.9.	Secondary Island, 2017. © *Kevin Ireland.*

Acknowledgements

This book was mostly written during the 2020–21 societal lockdown enforced in the UK during the Covid-19 pandemic. Having been working entirely from home for over a year, it is only as we put the final touches to our manuscript in August 2021 that restrictions are lifting and some sense of normality returns to everyday life. Working under such unusual circumstances has meant that we have been much more reliant on the generosity of others to provide insights, documents and images that we would normally not had to trouble them for, and we are grateful to the researchers, curators and archivists who have helped us assemble our book under these always surreal and, at times, difficult circumstances.

We would specifically like to thank Alex Ball, Robert Bell, Joe Cain, Nancy Chillingworth, Richard Fallon, Melvyn Harrison, Andrea Hart, Jonathan Jackson, Chris Manias, Ilja Nieuwland, Beau the greyhound, Darren Naish, Penny Read, Jim Secord, Helen Walasek, for assistance with images and discussions of the Geological Court. Lizzi Hewitt-Brown, Mollie Lyon, Lydia Lee at Bromley Council, Ed Morton of the Morton Partnership, David Carrington of Skillington Workshop, Ian Harper, Simon Buteux, Verena McCaig and Claire Brady at Historic England, and Teresa Heady are thanked for their insights and expertise on the maintenance and conservation of the Geological Court. Friends of Crystal Palace Dinosaurs members Sarah Jayne Slaughter, Jon Todd, Jeremy Young are sincerely thanked for their assistance and expertise on all manners related to the Crystal Palace Dinosaurs and for useful feedback on our manuscript.

We are also grateful to James Balston, Rhys Griffin, David Hone, Kevin Ireland, Chris Redgrave, David Vallade, the Crystal Palace Foundation, London Metropolitan Archive, London Picture Archive, the Trustees of the Natural History Museum, Natural History Museum Library & Archives and Imaging Unit, Historic England Archives, Princeton University Art Museum, Wellcome Collection, and the Wisbech & Fenland Museum for allowing us to use images from their collections. We have also benefited from many organizations who have scanned historic literature – books, scientific papers, newspapers, magazines and so on – and placed them all in the public domain at their own expense. These unsung heroes include the Internet Archive, Biodiversity Heritage Library, Google Books, the Wellcome Collection, Wikimedia Commons and Project Gutenberg.

EM would like to thank her wingmen Sarah Slaughter, Jeremy Young and Jon Todd in conquering the unusual physical challenges of our regular expeditions to eyeball our stupendous monsters, as well as giving intellectual and strategic input on how we can hold them together. She'd also like to thank all the many friends of the Friends of CP Dinosaurs whose wide-eyed enthusiasm for these monuments to the history of science in a south London park have provided the motivation for this book.

Finally, MPW would especially like to thank five-times book widow Georgia, who once again has shown tremendous patience and understanding about her husband spending much of his time and energy fussing about things and events that happened long, long ago. He promises to have the fireplace sorted before she reads this.

Index

Note: Pages carrying relevant information in illustrations or tables when not accompanied by relevant text are indicated by italic numbers. Such material is not separately identified on pages where text mentions have already been indexed.

A
Albert, Prince Consort 23, 25, 27
allosauroids 120, 165
Altispinax dunkeri 119
ammonites 11, *40, 42*, 50, *105*
amphibious habits 93–94, 126, 136–137, 141, 151–152
anatomical correlation 30–32, 34–37, 106, 143, *150*, 151
ankylosaurs 112–116
Anning, Mary and Joseph 128, 133
Anoplotherium genus 38, 52, 55, 68, 73, 78, *160, 162,* 179
  as camel-like *84*, 86, 88–90
  past restoration 85
  regarded as semi-aquatic 83, 85, 86
  sexual dimorphism 85, 86, 89
Anoplotherium commune 11–12, 35, 38, *69,* 83, 84–87
'*Anoplotherium gracile*' (later *Xiphodon gracilis*) 11–12, 35, 59, *70,* 83, *84,* 88–90
  misplaced as a *Megaloceros* fawn 70, 84, *87,* 88
Ansted, David Thomas 13, *22,* 25–26, 41–42, *46, 168,* 170
Anthropological Exhibition of Moscow 167–168
anti-evolutionists 27, 29–30, 116, 163, 173
antlers, fossils 60, 69–70, *71,* 176
aquatic species
  Anoplotherium as semi-aquatic 83, 85, 87
  Dicynodon as semi-aquatic 152
  '*Labyrinthodon*'/ *Mastodonsaurus* 147–148
  Teleosaurus 121–125
Aulacephalodon 150, 153
aurochs 27

B
Barilium dawsoni 107–110
Baxter, George *25,* 55–56
behavioural depictions 38, 76, 86, *87,* 90, 111
bipedality 87, 109, 111–112, 120, 163, 166–167, *169,* 171
  see also posing
British Museum (Natural History) *see* Natural History Museum
Bromley, London Borough 19, *177,* 180
Bromsgrovia walkeri 143, 148
Brunel, Isambard Kingdom 45, 63
Buckland, William
  Hawkins and 27
  honoured at the New Year banquet 64

Ichthyosaur
  reconstructions 128–130
  on plesiosaurs 133–134, *135*
  pterosaurs as fiendish 98
  see also Megalosaurus; *Pterodactylus*; '*Rhamphocephalus bucklandi*'

C
Campbell, James 13, 26, 46–47
Carboniferous Period 37–49, *44–45,* 141, 148
  see also Coal Measures
Centrosaurus apertus 106
Cetiosaurus oxoniensis 120
Chalk
  in the Geological Illustrations 51, 52, *54–55*
  Mosasaurus from 92
  pterosaurs perching 60, 91, 96, 98, 100
  see also Cretaceous Period
chimeric assemblages 35, 67, 107–109, *110,* 113, 143, 147–148
Chirotherium tracks 50, 142, 145–146, 147–148
Cimoliopterus cuvieri (previously '*Pterodactylus*' *cuvieri*) 11–12, 33, *54,* 60, 96, 99–102
classification *see* taxonomic …
clay moulds 18, 55, 57–58, *59,* 61, 63–64, 109
Coal Measures *40, 42, 45, 47,* 48–50, 70, *176,* 178
'*Colossochelys*' (now *Megalochelys*) 11, 55, 160
colouration 61, 62, 70–73, 79, 112, 116, 118
  '*Labyrinthodon*' sculptures 141–142
  pterosaur sculptures 97–98
  see also painting and paint analyses
commercialisation of enlightenment 17
communicating science to the public 17–18
comparative anatomy 29, *30,* 34–36, 67
  see also anatomical correlation
Concavenator corcovatus 119
conservation efforts
  Cliveden Conservation and Skillington Workshop 177–180
  Historic England and 19, 173
  historical timeline *177*
  Hylaeosaurus head 112
  on *Megalosaurus* 20
  Morton Partnership 46, *48, 177,* 179
  as poorly documented 18–19, 175
  as questionable 70
  as sporadic or delayed 19, 175
  supervised by Swinton *177, 178, 179*
Cope, Edward Drinker 94, *169*

Cretaceous period *162, 169*
  Geological Illustrations 50, 52
  Iguanodon and *Hylaeosaurus* 104, 107, 112
  pterosaurs 39, 52, 86, 100, *102,* 139
  see also Mosasaurus
crocodylians
  armour 38, 123
  gharials 76, 121–122, 125
  '*Labyrinthodon*' and 148
  mosasaurs mis-identified as 91–92, 94, *95*
  teeth 120, 123
  Teleosaurus as resembling 122–125
the Crystal Palace (building)
  fire of November 1936 8, 173
  rebuilt in Penge 9, 23, *24–25*
  temporary building in Hyde Park 21, 22
the Crystal Palace Company
  financial difficulties 45–46, 161–163, *177,* 178
  foundation and objectives 23–25, 180
  promotional activities 36, 62–63, 65
  recruitment of experts 25–32
  rejection of interpretation 24–25, 45
  Richard Owen and 29–32, 36, 119
Crystal Palace Park
  fountains 8, *10,* 23, 32, 45–46
  surviving dinosaurs 8
  see also Geological Court
Crystal Palace Park Trust 180
ctenosauriscids *12, 146,* 148
Cuvier, Georges
  on anatomical correlation 34–35
  on *Anoplotherium 84,* 85–86, 88–89
  on extinct mammals 68, 70–71, 74–76, 79, 82, 88
  on *Megatherium* 74
  on *Mosasaurus* 93
  on *Plesiosaurus* 134
  see also Cimoliopterus; '*Pterodactylus*' *cuvieri*
cycads 50, *52,* 110–111, 178

D
Darwin, Charles
  defended by Huxley 103
  views opposed by Hawkins 27, 163, 173
  views opposed by Owen 29–31
Deep Time concept 7, 41–42, 156, 183
deer *see Megaloceros*
Delamotte, Philip Henry *18, 45, 54,* 55, 60–61
dentition 90, 93, 99, *103, 108,* 115, *117, 143,* 145
Dicynodon
  building the sculpture 32, *59,* 93
  deductions from a single vertebra 36

models 57
  reconstruction as conjectural 93, 151–152
  on the Secondary Island 67, 141, 149–153, *172*
Dicynodon lacerticeps 11, 12, 149–153, *150–151,* 152
Dicynodon strigiceps 11, 12, 18, 150–153
diet, from gut contents 139
Dimorphodon 99
Dinomania 62, 157–158
Dinornis (moa) 30, *31,* 32, 36, 65, 143
dinosaur reconstructions
  as anatomically convincing 36–37
  predating CP sculptures 14
  as rhinocerine 30
dinosaur sculptures *see* sculptures, palaeontological
dinosaurs
  applying anatomical correlation 36
  coinage of the term 29, 104
  depictions in palaeoart 15, 104–106, 110–111
  first species named 104
  representation as mammalian-like 15, 104–106, 110
Doyle, Peter 25, *40, 41, 46,* 179

E
Earth, age of 16, 42
Earth Sciences *see* geology
educational role
  of the Crystal Palace enterprise 17, 24, 173
  of journalists 62
  of palaeoart 16, 18, 166, 171
elk, Irish *see Megaloceros*
enaliosaurs (marine reptiles) *12, 54,* 126–140, *159, 172, 176, 180, 183*
  ichthyosaurs as 127–133
  plesiosaurs as 133–140
  regarded as amphibious 126, 136–137
  see also ichthyosaurs; *Plesiosaurus*
evolution
  anti-evolutionists 27, 29–30, 116, 163, 173
  and politics 30
extinction, idea of 16, 68, 165

F
flint *40, 42,* 46, 49, *51,* 52–53
flying reptiles *see* pterosaurs
footprints *see* fossil trackways
fossil reconstructions 35, 74
  chimeras from 35, 107, 113, 143
fossil trackways 50, 142, *143,* 145–146, 147–148
fountains 8, *10,* 23, 32, 45–46
Friends of Crystal Palace Dinosaurs charity 19–20, 96, *177,* 180–183

G

Geological Column 41, *43*, 46
the Geological Court
    conservation efforts 13, 155
    construction 32–38, 39
    construction of the sculptures 25–32, 53–65
    cost overrun 39, 161
    delayed opening 39
    as last remnant of Crystal Palace Park 19
    original layout 10, 34
    as early palaeoart 34
    planned extent 13
    public and press reaction 156–158
    recommended order of viewing 67
    recruitment of specialists 25–32
    scope and ambition of the project 21
    sudden termination of the project 161–162
    viewed as an engineering project 21, 25
    see also sculptures, palaeontological
Geological Illustrations
    conformable arrangements 44–45
    construction and aims 39–41, *40, 42*
    guidebook omission *33*, 41
    industrial applications and 41–43, 46–47, 49
    lead mine *40*, 41, *42*, 46–48, 178
    Primary strata *42*, 46–49
    restoration efforts 41, *51*, 176
    Secondary strata 50–52
    Tertiary strata 52–53
    as unfinished 46, 65, 160
geology
    British geological strata 32, 41
    building public awareness 17, 26–27
Geology and Inhabitants of the Ancient World, by Owen 30, 32, 67, 91, 139, 144, 148, 150
gharials 76, 121–122, 125
Glossotherium 75
Great Exhibition of 1851 21, *22*, 23, 27
Greensand *11–12*, 52, 160, 177
Guide to the Crystal Palace and Park, by Samuel Phillips 24, *33*, 45, 49, 67, 83
guidebooks
    missing sculptures included 65
    purchase as essential 25, 32
    Routledge's 41, 47, 49, 63–64, 67, 70, 88
    see also Geology and Inhabitants

H

Hadrosaurus 111, *162*, 163, 165, *166*, 167, 169
Hastings Beds 107–109
Hawkins, Benjamin Waterhouse
    anatomical construction and posing 37–38, 55
    anti-Darwinian views 27, 163, 173
    College of New Jersey project *162*, 163, *166*, 171
    example drawings *28–29, 56*
    Great Exhibition aurochs 27
    influence of living species 38, 110, 147
    later career 162–163
    poster series on extinct animals *159*, 160, *168*, 170
    responsibility for the sculptures 13, *26*, 26–27, 32–34
    workshed engraving and reports *18*, 38, 56–59, *58*, 62–65, 88–89, 127, 134
Historic England 19, 173, *177*, 179–180
Hylaeosaurus armatus
    conservation work 177
    under construction 61
    in Hawkins' poster series 160
    images of *12, 18, 35, 54–55, 57, 105–106, 162, 182*
    Mantell's influence 29
    modern interpretations compared 106, 115–116
    nature of armour 114
    in palaeoart 15, 160
    in recognition of Dinosauria 30, 104
    sculpture *11–12*, 112–116, *182*
Hypselospinus fittoni 109

I

ichthyosaurs
    early palaeoart examples 126, 133
    illustrations and depictions *15, 105, 127, 131–132, 167*
    Jurassic Lias and 50
    reference fossils 124, *129*
    representation and taxonomy 127–133
Ichthyosaurus communis *11–12, 45, 54, 57*, 60, 126–127, *128–129*, 130–131, *176, 180*
'*Ichthyosaurus*' (now *Leptonectes*) *tenuirostris* *11–12, 54, 59*, 60, 127, *128–129*, 130–132
'*Ichthyosaurus*' (now *Temnodontosaurus*) *platydon* *11–12, 54*, 127–128, *129*, 130, 132–133
Iguanodon
    Anthropological Exhibition of Moscow 167–168
    Geological Court display *17, 37, 105–106, 176, 182*
    Hawkins' inferences 36
    Kuwasseg artwork of *15, 167–169*, 170
    nasal horn controversy 32, *108*, 109–110
    in palaeoart 15, 160, 169
    in recognition of Dinosauria 30
    Wisbech model 57
'*Iguanodon anglicus*' 108
Iguanodon bernissartensis 107, 112, 163, *170*
'*Iguanodon mantelli*' *11–12*, 104, *107–108*, 107–112
    complete skeletons 112
    reference fossils 107–109
    sculpture 107–112
    see also Mantellisaurus
Iguanodon mould
    in Hawkins' workshed 18
    New Year's Eve banquet, 1853 18, 31, 63–65, 83
interpretative material 16, 24–25, 45, 165

J

Jurassic Coast, Dorset 14, 50, 126, 128, 132, 134, 137
Jurassic Park (film) 17, 174

K

Karoo Supergroup 149
Knight, Charles 172
Kuwasseg, Josef *15*, 143–144, *167–169*, 168, 170

L

'*Labyrinthodon*' 32, *33*, 36–37, *58*, 144–145
    regarded as amphibian 141–142
'*Labyrinthodon pachygnathus*'
    deductions about 36–38
    depictions and illustrations *12, 18, 33, 54, 58–59*, 88, *143, 162, 169, 172*
    Marsh's view of 166
    models and derived illustrations *158*, 159, 170
    Owen's view of 143–145
    on the Secondary Island *54*, 67, 141–146, 148
'*Labyrinthodon salamandroides*' 12, 141, *142–143*, 147–148
'*Laelaps*' (*Dryptosaurus*) *162*, 163, 165–167, *169*
lead mine feature *40*, 41, *42*, 46–48, 178
Leptonectes (previously '*Ichthyosaurus*') *tenuirostris* 12, 127–132, *180*
Listed Building status 19, 173, *177*, 179–180
llamas 84, 88–90

M

Macrospondylus bollensis (formerly '*Teleosaurus chapman(n)I*') *12, 121*, 122–123, 125
mammoths 34, *55, 57*, 65, 69, 160–161, 167
Mantell, Gideon *26*
    on anatomical correlation 35
    association with Hawkins 27
    association with Owen 30–32
    declined Covent Garden Company's invitation 27–29
    on *Hylaeosaurus* 112–114
    on *Iguanodon* 27, 109–111, 163
    on *Mosasaurus* 93
    nasal horn controversy 32, *108*, 109–110
    see also '*Iguanodon mantelli*'; *Mantellisaurus*
Mantellisaurus atherfieldensis 107–109
marine reptiles *see* enaliosaurs; *Macrospondylus*; *Mosasaurus*
Marsh, Othniel C 94, 100, 165–166, 171, 173–174
Martin, John *15*, 98, 104, *105*, 126, 133
mastodons 34, 63, 69, 142, 160
Mastodonsaurus giganteus *12*, 142, *143*, 145, 147–148
'*Megaceros*' *hibernicus* *11–12*, 71
    see also Megaloceros
Megaloceros giganteus 53, 69–73
    early fossils *16*, 34
    fawn sculpture 70, 84, *87*, 88
Megalochelys (formerly '*Colossochelys*') *11*, 55, 160
Megalosaurus bucklandii
    damage and repair 20, 47, *176, 181*
    envisaged lifestyle 19, *106*
    Hawkins' inferences 36, *118*
    in palaeoart 15, 160, 169
    recognition of Dinosauria 30
    sculpture 61, 62, 116–120, *182*
Megatherium americanum
    depictions and illustrations *11–12, 29, 58–59, 160, 162, 166*
    near-complete fossils 34
    on the Tertiary Island 68, 73–77, 178
    unique construction 60
moa 30, *31*, 32, 36, 65, 143, 160
models, full-size *see* sculptures
models, Tennant's 158–160
models, Wisbech and Tennant's 25, 55–56, *57*, 158–160
Mosasaurus hoffmani
    deduced anatomy 36, 38, 93–95
    fossil remains *92*–93
    in Hawkins' workshop 59
    location of the sculpture 11–12
    previous restorations 11–12, 32, *164*, 176
    regarded as amphibious 93–94
    sculpture of 91–95
'*Mosasaurus maximiliani*' (now *M. missouriensis*) 94
Mosasaurus missouriensis *92*, 94
Mountain Limestone illustration *40, 42*, 47–49, 178
musculature, depictions 36–37, 57, 76, 86, 101, 110–111, 118–119
museums *see* interpretative material; Natural History Museum
Mylodon robustus 75–76

N

national biases 21, 158, 166
Natural History Museum (previously British Museum (Natural History)) 23, 29, 116, 138, 178, *179*
New Jersey, College of (now Princeton) *162*, 163, *166*, 171
New Red Sandstone *40, 42*, 45, 50, 52, 141–142, 151
New Year's Eve banquet, 1853 *18*, 31, 63, 83
nodosaurids 112–114

O

Old Red Sandstone *40, 43*, 46–47
Oolite 9, *11–12, 40*, 50, *51*, 97, 102–103, 122, *176*, 177, 179
Owen, Richard *26*
    as an authority on mammals 68
    demonization 31
    Geology and Inhabitants of the Ancient World 30, 32, 63, 67, 91, 118–119, 136, 139, 144, 148, 150
    on ichthyosaurs 128–130
    on '*Labyrinthodon*' 142, 144, 148
    and the Natural History Museum 29, 116
    at the New Year's Eve banquet, 1853 63
    as nominal consultant 13, 27, 29–32, 38, 55–56
    opinions on evolution 30
    on plesiosaurs *135*, 139

proposed dinosaur body plan 106
on pterosaurs as fiendish 98
publications contradicting sculptures 19, 110
'sauroid batrachians' 142–143, 147–148
on teleosaurs 122, 123
tribute to Hawkins 155

P
'Pachydermata' 82
painting and paint analyses 62, 70–71, 73, 79, 85, 97, 110, 112, 116, 118, 175
palaeoart
complexity of 19
critique of and opposition to 164–166
Geological Court as pioneering 14–15, 19, 36–38, 156, 168–170, 172–174
living animals influencing 34, 37–38, 75, 79, 99, 118, 135, 136, 151
public educational role 16, 18, 166
see also Hawkins, Waterhouse; sculptures, palaeontological
Palaeotherium genus
depictions and illustrations 18, 35, 54–55, 58–59, 69, 80–81, 160, 162
moving into position 60, 85
tapir resemblance 37, 68, 79, 87
Palaeotherium magnum 81–83
as a missing sculpture 12, 19, 77–79, 81–83, 177
modern interpretation 83
Palaeotherium medium 12, 37, 77, 79–80
'*Palaeotherium minus*' (now *Plagiolophus minor*) 12, 35, 55, 77–79, 79, 80, 81, 177
Paris Basin 34, 35, 53, 79, 85, 88–89, 160
Paxton, Sir Joseph
designer of the first Crystal Palace 21, 22
involvement with the Geological Court 25–26, 32, 45–46
layout of the park and gardens 171
urging Crystal Palace rebuild 23
Pearson, Helga 152
Peck, Robert M 27, 63, 173
Penge (Sydenham Hill) site 9, 23, 39, 45
Petrifactions and their teachings... by GA Mantell 29
petting zoo (Pets Corner) 78–79, 85, 89, 177–178
Phillips, Samuel 7, 24–26, 67
guidebook 24, 33, 45, 49, 67, 83
photogrammetry 182
Plagiolophus minor (previously '*Palaeotherium minus*') 12, 35, 55, 77–78, 79, 81, 177
Pleistocene epoch 35, 69–73, 160, 162
Plesiosaurus (genus) 63, 98, 126, 133–140, 158, 162
Plesiosaurus dolichodeirus 12, 54, 59, 60, 127, 134, 139

'*Plesiosaurus*' *hawkinsii* (now *Thalassiodracon hawkinsii*) 12, 54, 127, 134, 140, 178
'*Plesiosaurus*' *macrocephalus* 11–12, 32, 137–139
depictions and illustrations 45, 78, 134, 158–159, 180
head lost and replaced 177–178
wrongly assigned to the genus 138–139
pliosaurids 139–140
Plues, Margaret 168, 170
posing
Anoplotherium 86–87
ichthyosaurs 131–132, 133
Iguanodon 111–112
Megatherium 73–76
Palaeotherium 79
plesiosaurs 134–136, 139
pterosaurs 97, 101, 103
Teleosaurus 121
poster series 159, 160, 168, 170
Primary strata, Geological Illustrations 42, 46–49
proboscides 76, 80–83
see also trunks
Protocupressinoxylon purbeckensis 50, 51
pseudosuchians 146–148
Pteranodon 99, 100
pterosaurs 96–103
Chalk pterosaurs 98, 100–102, 176
constructional challenges 36, 60
Dimorphodon 99
diversity 99
Dorygnathus 103
early palaeoart examples 126, 127
Klobiodon rocheri 103
missing Jurassic sculptures 96
Oolitic pterosaurs 97, 99, 102–103
Pteranodon 99, 100
Rhamphocephalus bucklandi (formerly *Pterodactylus bucklandi*) 97, 99, 102–103
Pterodactylus antiquus 96, 98–100, 103
'*Pterodactylus bucklandi*' (now '*Rhamphocephalus bucklandi*') 11–12
'*Pterodactylus*' *cuvieri* (now *Cimoliopterus cuvieri*) 11–12, 33, 54, 60, 96, 99–102
public engagement with geology 16–17, 19, 26–27, 165

R
'*Regnosaurus northamptoni*' 113–114
'*Rhamphocephalus bucklandi*' (formerly '*Pterodactylus bucklandi*') 97, 99, 102–103
rhomaleosaurids 133, 138–139

S
Scelidosaurus harrisonii 116, 165, 171
science, fusion with artistry 17
science communication and contemporary understanding 17–18

sclerotic rings 130
sculptures, palaeontological
construction process and cost 53–65, 73, 109–110
Cretaceous dinosaurs 107–116
effect of later fossil finds 111–112, 165
engineering approach 53, 134–136, 138
Listed Building status 19
of mammals 68–90
of marine reptiles 91–95, 126–140
not assembled *in situ* 60
number of real dinosaurs 10
numbers originally built 18, 68
people involved in the construction 25–32
placed before geological features 46
recommended order of viewing 67, 69
selection criteria for animals 34
Tennant's and Wisbech models 25, 55–56, 57, 158–160
Secondary Island 13, 44, 47, 50, 51, 52, 54, 172
guidebook coverage 32, 33
Secondary Island bridge 19, 91, 177, 180, 181
sloths, ground *see Megatherium*; *Mylodon*
South Kensington museums and colleges 23
spinosaurids 120
stegosaurs 113–114, 169
stem-mammals 149, 153
Steneosaurus 122
Stonesfield Slate 99, 102–103
stratigraphy 41, 44
superposition, geological 41–42
Sydenham Hill site, Penge 9, 23, 39, 45

T
taxonomic confusion
Ichthyosaur reclassification 127
over *Iguanodon* 107, 112
over teleosaurs 122
wastebasket taxa 99, 107, 118, 150
taxonomic history of *Megalosaurus* 71
Teleosaurus cadamensis 122
'*Teleosaurus chapman(n)i*' (now *Macrospondylus bollensis*) 121–125
armour 38
construction 60
depictions and illustrations 9, 11–13, 54, 59, 159, 162, 172
deterioration and conservation 176, 178, 179, 183
Temnodontosaurus 130, 133, 158, 162
repairs to sculpture 127–128, 176
Temnodontosaurus platyodon 12, 126–127, 128, 129, 132
Temnospondyli 12, 142–143, 147–148

Tertiary Epoch revised 53
Tertiary Island
guidebook omissions 32, 68
as location of mammalian sculptures 55, 65, 68
never completed 39, 46, 52–53
rock types 44
Thalassiodracon hawkinsii see '*Plesiosaurus*' *hawkinsii*
Thalassiodracon hawkinsii (formerly '*Plesiosaurus*' *hawkinsii*) 12, 54, 127, 134, 140, 178
Thalattosuchia 122
Thaumatodracon wiedenrothi 138
Tidal Lake
aquatic species 83, 84, 86, 121
features revealed 50, 91, 92, 126
planned operation 45–47, 177
Titanites ammonites 42, 50
tortoises *see* turtles
Torvosaurus 120
tree fossils 50
trunks (animal) 60, 75–77, 79–83, 134
see also proboscides
turtles
Colossochelys (now *Megalochelys*) 11, 55, 160
ichthyosaurs and 130
as model for *Dicynodon* 36, 141, 150–153
as model for *Teleosaurus* 123
Tyrannosaurus 119

U
unconformities, simulated 45, 50
unicornum verum 35
United States, fossil finds 92, 94, 100, 111, 116, 158, 167, 171

V
vandalism 20, 48, 50, 77, 79, 81, 89, 97, 175, 176
Victorian Age
monsterizing tendency 86, 158–159
palaeoart 122, 126, 183
understanding of prehistory 19, 141

W
waterfall 49
Wealden deposits
dinosaurs of 104, 107–116, 118, 168
in the geological column 43
in Geological Illustrations 50, 51–52
location 11–12, 40
Wealden supergroup 107, 112, 118–119
weathering 69, 175
Whitby Mudstone 122
Wisbech (& Fenland) Museum models 25, 55–56, 57

X
Xiphodon gracilis (previously '*Anoplotherium gracile*') 11, 12, 55, 84, 87–90